The Shape and Size of the Earth

Dino Boccaletti

The Shape and Size of the Earth

A Historical Journey from Homer to Artificial Satellites

Springer

Dino Boccaletti
Rome
Italy

ISBN 978-3-030-08050-1 ISBN 978-3-319-90593-8 (eBook)
https://doi.org/10.1007/978-3-319-90593-8

Softcover re-print of the Hardcover 1st edition 2019

Printed on acid-free paper

This Springer imprint is published by the registered company Springer International Publishing AG part of Springer Nature
The registered company address is: Gewerbestrasse 11, 6330 Cham, Switzerland

To Ruben and Matilde

Preface

How an Interest was Born, Developed, and Produced a Book

The reader may be intrigued by the genesis of a book such as this. Actually, what stimulated me to think about this subject was another interesting reading.

Many years ago, in the first phase of my teaching in the University of Rome "La Sapienza", I was charged with teaching a course on Institutions of Mathematics to students pursuing a degree in Chemistry. In that occasion, I happened to read a very beautiful book: *Mathematical Methods in Science* of George Pólya, in which I found this passage:

> That Eratosthenes' result is inaccurate does not really detract from the greatness of his achievement. It is his method that excites our admiration. Would not a giant measure the Earth by encircling it with his arms to compare its circumference with his span? And what did our little pygmy Eratosthenes do? At Alexandria at noon on a certain midsummer's day long ago, he observed the shadow cast by a little stick and used his protractor. A mere shadow and the pygmy a giant who spanned the Earth.[1]

The reading of this passage, addressed to the measurement of the terrestrial meridian executed by Eratosthenes (in the third century BC), of which until then I had a superficial knowledge, and the appreciation—enthusiastic to say the least—that Pólya attributed to that event, inspired my interest in expanding on the subject, and I started a series of considerations.

First of all, I began to think that, through the centuries, the great significance attributed to the Copernican Revolution ended up obscuring another great conquest of the human thought: the sphericity of the Earth. The idea and personal conviction, contrary to the immediate evidence, that the ground on which one was walking was not a roughly flat surface (apart from the mountains!), but was rather the surface of a sphere whose boundaries could not be discerned, was undoubtedly a great conquest (even if it could not be defined as a revolution). Earlier on, one confused the sphere with the tangent plane!

---

[1] G. Pólya: *Mathematical Methods in Science* (Washington, 1977).

Pólya's admiration for Eratosthenes was based on the method he had used: Eratosthenes had, so to speak, measured the terrestrial meridian while seated in his room and using mathematics. From this, the power of mathematics is evident, as was underlined by Eugene Wigner some years before in his celebrated paper "The Unreasonable Effectiveness of Mathematics in the Natural Sciences."[2]

At the beginning of the paper, Wigner used the rhetorical expedient of narrating the conversation between two old high school classmates, one of whom had become a statistician working in population trends. This one showed to the other a reprint of a publication of his where a Gaussian distribution and also the symbol π appeared. The friend asked:

> "And what is this symbol here?" "Oh", said the statistician, "This is pi". "What is that?" "The ratio of the circumference of the circle to its diameter." "Well, now you are pushing your joke too far," said the classmate, "surely the population has nothing to do with the circumference of the circle."

Here, we are at the root of the problem of interpreting nature with the tool of mathematics and of the dual nature of mathematics as both a tool and an independent science: a problem that is up to us to deal with. We only want to emphasize the maieutic role that some problems presented by nature have had in giving birth to branches of mathematics during history. Of course, as all know, three centuries before Wigner, Galileo (in the *Assayer*, 1623) had upheld the indispensability of mathematics for interpreting nature. If we want go back further, even Robert Grosseteste (see Sect. 3.5.2) had upheld the same.

The second consideration is this: once established that it had been demonstrated that the Earth had a spherical shape since the middle of the third century BC, and that an evaluation of its size had also been proposed, how did the thought on this question evolve in the following centuries? Had the fact that an adequately rigorous answer to such an important question had been given since the third century been sufficient to convince the general opinion? Actually, quoting the result of a discovery (or of a theory), such as that of Eratosthenes (as well as those of the philosophers who preceded him), means establishing only its chronological priority, but not the moment when this discovery (or theory) has become part of the cultural heritage of humanity.

An investigation of the problem leads to the discovery that the notion of the Earth's sphericity, which nowadays the children learn since the primary school, took centuries to be universally accepted. As happened with other questions, the Catholic Church also took part in the discussion, which lasted for centuries, together with all the philosophers of the Middle Ages. The debate arose from the conflict between what was written in the Bible and the results of the natural sciences.

---

[2]E. Wigner, "The Unreasonable Effectiveness of Mathematics in the Natural Sciences", in *Communications in Pure and Applied Mathematics* 13(1) 1960: 1–14.

Obviously, during the centuries, the scientific research evolved, even arriving to the point of stating that the reference figure is no longer a sphere (it has been called a "geoid") and that, due to its rotation, the shape is oblate (through the disputations and the measurements occurred in the eighteenth century). Through the work of Newton, Clairaut, Maupertuis, Delambre, and others, and through the progress of technology, we have arrived to satellite geodesy. Every stage of these both theoretical and experimental advancements has constantly involved the development of new mathematical techniques, always reinforcing Wigner's assertion.

Narrating this fascinating story from the beginning reveals more about the evolution of the human thought than only about the evolution of the specific concept of Earth's sphericity, and is certainly worth the effort.

In fact, there are many works which deal with the subject, many of which mainly concern the controversies on the Earth's ellipsoid shape,[3] but the ambition of this present book is that of pointing out how from the very beginning the speculations about the shape and the size of the Earth were not confined to the technology but traversed the evolution of human thought itself.

Notice to the Reader

Subjects The reader may wonder at not finding certain subjects dealt with when particular authors or particular periods are mentioned. We have set ourselves to cover exclusively what concerns the theories and the experiments on the shape and the size of the Earth and, as said above, their impact on secular society. Thus, for instance, when we dealt with Ptolemy's *Geography* and projections therein, we did not seize the opportunity to talk about cartography despite the fact that it was an appealing subject given its relationship with mathematics.

Excerpts Except for the Greek philosophers, we have always preferred to supply the text with English translations of works being discussed and report in footnotes the quoted excerpts in their original language. This is done to avoid arbitrary interpretations and variants of tradition in subsequent quotations. We have not always found the chosen Latin excerpts translated into English by competent scholars. In those cases, we have compensated for lack with a translation of our own. Let us hope we have obtained a passable result.

Bibliography Every single chapter is equipped with footnotes referencing papers and books quoted. We considered them exhaustive and decided not to weight each chapter down with the usual ad hoc bibliography. We have rather preferred to include a list of suggested readings, often classical textbooks, for a prospective widening of the context.

---

[3] Suffice it to mention the classic *A History of the Mathematical Theories of Attraction and the Figure of the Earth* by Isaac Todhunter (London: Macmillan, 1873) and the modern *The Problem of the Earth's Shape from Newton to Clairaut* by John L. Greenberg (Cambridge University Press, 1995).

Acknowledgements

I wish to thank my wife Maria Grazia and my daughter Chiara for their invaluable help in drawing up and completing this book.

The excellent editing by Kim Williams is gratefully acknowledged.

Rome, Italy
February 2018

Dino Boccaletti

Contents

List of Figures

Chapter 1
The Graeco-Roman World

Today, the earth, on which we move every day from one way to another and build our houses, can be nominally transformed, by ourselves, with a simple change of type (lower-case → upper-case) into something more important and greater: the Earth. If we write it with a capital letter, it means that we are talking about the planet Earth,[1] that is, about a body which, together with the other planets of the solar system, can be called—following the ancient locution—a celestial body. This transformation (ideal and formal), which we effect by using the same noun but passing from the lower-case to the upper-case, has taken at least seventeen centuries to become the common heritage of humanity (from the time when the problem was first faced in terms of scientific research, with Aristarchus of Samos, until the time of Copernicus).

What we are talking about, and about which we shall talk more later, is to be understood in reference to the Western civilization. Undoubtedly, to the best of our knowledge, the problem was also faced and dealt with in the preceding civilizations (Egyptian, Babylonian, etc.)[2] but always at a mythical-religious level, according to the scanty extant documents. To be fair, the extant documents are also very scanty with regard to Western civilization, i.e., that exceptional laboratory of thought which was the Greek world since the seventh century BC.[3] Before that date, one has to do with narrations which belong to the world of mythology. Before the transition from *mýthos* to *lógos* (to use the terms dear to philosophers) natural phenomena were credited to gods, and the world of gods was very crowded, since they were

[1]From now on we shall always use the upper case (except in the quoted excerpts, respecting their original form).

[2]It seems that the Babylonians had conceived the Earth as a disc with a central mountain encircled by the ocean and with a system of mountains which supported a solid hemispheric sky.

[3]As the great German scholar Bruno Snell says in the introduction of his fundamental work *Die Entdeckung des Geistes. Studien zur Entstehung des europäischen Denkens bei den Griechen* (1948), the thought in its logical forms common to us Europeans arose among the Greeks, and indeed since that time it has been considered the only possible form of thought.

D. Boccaletti, *The Shape and Size of the Earth*,
https://doi.org/10.1007/978-3-319-90593-8_1

presumed to be responsible for every natural phenomenon (wind, rain, earthquakes, etc.) and had to act as the protectors of various professional activities as well.

The most important singers of the world of gods, were, in chronological order, Homer and Hesiod. Obviously, with regard to Homer, we are not interested in the so-called "Homeric question" (i.e., the discussion of whether a poet bearing this name and author of the two poems *Iliad* and *Odyssey* really existed), but only in what was written in those two works (before the seventh century BC) regarding our theme: the shape and size of the Earth.

1.1 Homer and Hesiod

The Greek world in the time of Homer was much wider than the present Greece and, although most of the movements had to take place by sea (given the great number of the islands), the Greeks traveled a lot. There is a trace of this in Homer's poems, where various places of the Mediterranean and other beyond the so-called Pillars of Heracles are quoted and described. This abundance of descriptions of different places even led Strabo (eight centuries later) to define Homer the founder of the science of geography[4] (we shall come back to this subject later).

Before looking at the passage of the *Iliad* which concerns our theme, we must point out an important circumstance. As we have said, the considerations on the Earth (shape, size, etc.) began relatively late. Instead, the considerations on the sky, for both religious and practical reasons, began many centuries before. Men had realized relatively early that the motions of celestial bodies displayed periodicities and influenced the life which took place on the Earth. The Babylonians had also developed rules for deducing instructions and forecasts from the motions of celestial bodies (i.e., astrology). It goes without saying that it was considered more important to deal with the sky than with the Earth. Therefore, the interest in the Earth was confined to its insertion in a cosmography in which, obviously, the predominant element was the celestial sphere.

In fact, this is what we find in the *Iliad*, i.e., precisely a model of the Homeric universe. In Book XVIII, devoted to the making of the arms for Achilles (his mother Tethis applies to Hephaestus for this work), the shield is described in minute detail and we can grasp it as a concise representation of the macrocosm and the microcosm. The Earth is a disc bordered by river Oceanus (the border of the disc represents the frontiers of the world). Above it, there is the heavenly vault, and the whole (Earth, sea, sky) portrays the Homeric view of the world. The description of the shield occupies more than a hundred verses (474–607) of Book XVIII. Let us see those where the Homeric description of the world is established:

> He wrought the earth, the heavens, and the sea; the moon also at her full and the untiring sun, with all the signs that glorify the face of heaven—the Pleiads, the Hyads, huge Orion,

---

[4]Strabo: *Geography*, vol. I (Eng. trans. Horace Leonard Jones).

> and the Bear, which men also call the Wain and which turns round ever in one place, facing Orion, and alone never dips into the stream of Oceanus.
>
> …
>
> All round the outermost rim of the shield he set the mighty stream of the river Oceanus.[5]

Hesiod, who is the first Greek poet whose existence is historically testified (end of the sixth century BC), also moves in the field of mythology. In his work *Theogony*, he tells the history of the succeeding generations of the gods, who originate from the sky and the Earth. There too the Earth (Gaia) appears as a protagonist:

> From Chaos came forth Erebus and black Night; but of Night were born Aether (5) and Day, whom she conceived and bare from union in love with Erebus. And Earth first bare starry Heaven, equal to herself, to cover her on every side, and to be an ever-sure abiding-place for the blessed gods.[6]

Further on, Earth will be involved in the bloody struggles between Uranus and Cronos and in the succeeding generations.

If in the *Theogony* there are no references to the nature of the Earth and its shape, these are also lacking in the Hesiod's other work, *The Works and the Days*, devoted to life on Earth in all its forms, man's work included. In contrast to Homer, who quotes various places in both Greece and neighbouring countries, Hesiod declares his sedentariness in addressing his brother Perses and saying that he has traveled by sea only one time, for a little trip from Aulis to Euboea.

Neither Homer nor Hesiod ever explicitly hint at the shape of the Earth. What we can deduce from this is either that the argument had not yet come to be of interest, or that the Earth was implicitly considered to be flat, as it will be for the earliest philosophers. While in Homer Achilles' shield lends corroboration to this theory (Earth flat like a disc bordered by river Oceanus), in Hesiod we can only say that the theory is not contradicted.

In the ensuing work of an unknown author, *The Shield of Heracles* (in ancient times credited to Hesiod), after the description of various events, at the end we read: "And round the rim Ocean was flowing, with a full stream as it seemed, and enclosed all the cunning work of the shield".[7]

1.2 The Ionic School

We can date the flourishing of a philosophical school on the Aegean coast of Asia Minor, the first one in chronological order, to the turn of the seventh century BC. The town of Miletus was in a position very accessible to receive external influences

---

[5]Homer: *Iliad*, XVIII, 483–489, 606–607 (trans. Samuel Butler).

[6]Hesiod: *Theogony* 116–138 (Eng. trans. Hugh G. Evelyn-White).

[7]*The Shield of Heracles* 314–316, trans. Hugh G. Evelyn-White.

from both East and West. The first philosopher (so he is handed down) and "founder" of the Ionic School is precisely Thales of Miletus (about 624–565 BC) who, according to his biographers, visited both Mesopotamia and Egypt, bringing from there knowledge of various kinds, mainly astronomical. We have talked of biographers, but the news regarding Thales (life and works) actually come from fragments[8] and quotations on the part of far subsequent philosophers and of the so-called doxographers (not always reliable) of whom the most important is Diogenes Laertius (third century AD!).[9]

As Bruno Snell remarks,[10] the contrast between mythical image and logical thought (mýthos → lógos) comes into sharp focus in the causal interpretation of nature. In this field, the transition from mythical to logical thought also acquires direct evidence: that which in the ancient times was interpreted as a work of gods, demons and heroes, will later be interpreted in a rational sense. The initiator of this way of thinking was Thales. We quote from Aristotle's *Metaphysics*:

> Of the first philosophers, then, most thought the principles which were of the nature of matter were the only principles of all things. That of which all things that are consist, the first from which they come to be, the last into which they are resolved (the substance remaining, but changing in its modifications), this they say is the element and this the principle of things, and therefore they think nothing is either generated or destroyed, since this sort of entity is always conserved.
>
> Yet they do not all agree as to the number and the nature of these principles. Thales, the founder of this type of philosophy, says the principle is water (for which reason he declared that the earth rests on water), getting the notion perhaps from seeing that the nutriment of all things is moist, and that heat itself is generated from the moist and kept alive by it (and that from which they come to be is a principle of all things). He got his notion from this fact, and from the fact that the seeds of all things have a moist nature, and that water is the origin of the nature of moist things.[11]

Therefore, according to Thales, the water is the unique principle of all things and also the Earth floats upon the water like a piece of wood.[12]

Thales (at least from what can be extracted from the quotations and the fragments) does not forge ahead. He does not say anything regarding what supports the water nor on a possible shape of the Earth. The important point for him is that of

---

[8]Among the numerous (incomplete) translations into English of Hermann Diels's *Die Fragmente der Vorsokratiker griechisch und deutsch* (Berlin, 1903), one can refer to Kathleen Freeman: *Ancilla to Pre-Socratic Philosophers* (Harvard University Press, 1948; rpt. 1983).

[9]Diogenes Laertius: *Lives of Eminent Philosophers*, 2 vols. trans. R.D. Hicks, Loeb Classical Library (Cambridge, MA: Harvard University Press, 1925).

[10]B. Snell, op. cit., Chap. XI, 5.

[11]Aristotle: *Metaphysics* I, 983 b6 (Eng. trans. W. D. Ross).

[12]Aristotle: *On the Heavens* II, 294a: "Others say the Earth rests upon water. This, indeed, is the oldest theory that has been preserved, and is attributed to Thales of Miletus. It was supposed to stay still because it floated like wood and other similar substances, which are so constituted as to rest upon but not upon air. As if the same account had not to be given of the water which carries the Earth as of the Earth itself!" (Eng. trans. J. L. Stocks).

having identified what is the fundamental element. As we shall see later on, a unique element will not be trusted by his successors.

Anaximander (610–546 BC) also belonged to the Ionic School. He was a contemporary and a fellow-citizen of Thales (and also a disciple of him). Anaximander broadened the philosophy of Ionians into a wider horizon: he envisaged the origin of all things coming from a unique substance which he called principle (*arché*), which lay not in water (as Thales maintained), in air or in any other element, but rather in the infinite (*apeiron*), from which all things originate and to which they return when they have ended their vital cycle. From the point of view we are interested in and of which we have to give account, Anaximander conceived of the Earth hanging in the air and not sustained by anything other than the symmetry principle (!). In fact, since in his opinion the heavens performed a complete revolution with a great vortex, the Earth stayed at the centre and, all directions around the centre being equivalent, the external vortex obliged the Earth to remain where it was. Aristotle says:

> … there are some, Anaximander, for instance, among the ancients, who say that the Earth keeps its place because of its indifference. Motion upward and downward and sideways were all, they thought, equally inappropriate to that which is set at the centre and indifferently related to every extreme point; and to move in contrary directions at the same time was impossible: so it must needs remain still.[13]

That is, there was not any reason because of which the Earth went in one direction rather than in another: in this way Anaximander introduced what we moderns call the sufficient-reason principle.[14]

With regard to the shape, the Earth "… the figure of [the Earth] is curved, circular, similar to a column of stone. And one of the surfaces we tread upon, but the other is opposite".[15] Anaximander was the first to draw a map of the known world which (so it seemed) was also reproduced on a bronze plate. Later on, the map was corrected by the Milesian Hecataeus (560–490 BC). As opposed to what held by the doxographers just quoted, Diogenes Laertius ascribes to Anaximander the idea of a spherical Earth[16] (which would, moreover, better conform to the theory

[13]Aristotle: *On the Heavens* II, 13, 295b 10 (Eng. trans. J. L. Stocks).

[14]See: Giorgio de Santillana: *The Origins of Scientific Thought: From Anaximander to Proclus* (Chicago: The University of Chicago Press, 1961), chap. 1.

[15]This was told by Hippolytus of Rome (175–235 AD), a Christian writer, in his work *Refutatio omnium haeresium*, Chap. V (see Hippolytus Romanus: *The Refutation Of All Heresies*, trans. J. H. MacMahon) and, before him, by Aëtius in *Placita Philosophorum* (III, 10).

[16]Diogenes Laertius, *Lives*, op. cit. book II, 1: "Anaximander the son of Praxiades, was a native of Miletus. He laid down as his principle and element that which is unlimited without defining it as air or water or anything else. He held that the parts undergo change, but the whole is unchangeable; that the Earth, which is of spherical shape, lies in the midst, occupying the place of a centre; that the moon, shining with borrowed light, derives its illumination from the sun; further, that the sun is as large as the Earth and consists of the purest fire".

of an Earth being balanced in the universe for symmetry reason). In any case, the discussion is still open.[17]

Diogenes Laertius, using the work of the philosopher Favorinus (85–143 AD) as a source, also credits Anaximander with the invention of the gnomon (which Herodotus, instead, attributes to the Babylonians[18]) as well as the calculation of the equinoxes.[19]

According to the history of philosophy, after Thales and Anaximander, the third outstanding philosopher of the so-called Ionic School is Anaximenes (586–528 BC) of Miletus, younger than Anaximander and perhaps a disciple of him. He extended Anaximander's thought, even if he came back to hold the existence of a fundamental element, in this case air. In his opinion air was the essence of all things and he assimilated it to the vital essence which he called *pneuma*. Also, the universe itself is thought as a living being. According to the well-known fragment (from Aëtius):

> Anaximenes his fellow-citizen pronounceth, that air is the principle of all beings; from it all receive their original, and into it all return. He affirms that our soul is nothing but air; it is that which constitutes and preserves; the whole world is invested with spirit and air.[20]

Here we are not so much interested in setting forth Anaximenes' philosophical thought as his idea regarding the Earth. Aristotle, in *On the Heavens*, relates:

> Anaximenes and Anaxagoras and Democritus give the flatness of the earth as the cause of its staying still. Thus, they say, it does not cut, but covers like a fid, the air beneath it. This seems to be the way of flat shaped bodies: for even the wind can scarcely move them because of their power of resistance. The same immobility, they say, is produced by the flatness of the surface which the Earth presents to the air which underlies it; while the air, not having room enough to change its place because it is underneath the Earth, stays there in a mass, like the water in the case of the water-clock. And they adduce an amount of evidence to prove that air, when cut off and at rest, can bear a considerable weight.[21]

It can be said that the philosophical research of the Ionic philosophers reached its high point with the doctrine of Heraclitus of Ephesus (flourished about 504–501 BC). Heraclitus is the philosopher of the becoming, of the never-ending change and also of the fire, which, according to him, is the principle of all things, and plays a completely different role from the fundamental elements theorized by Anaximander and Anaximenes. The philosophy of nature[22] of Heraclitus is no more directly tied

---

[17]See in this regard F. Enriques, G. De Santillana: *Storia del Pensiero Scientifico*, vol. I (Treves: Tumminelli, 1932), pp 265–270.

[18]Herodotus: *The Histories*, book II, 109: "For as touching the sun-dial and gnomon and the twelve divisions of the day, they were learnt by Hellenes from Babylonians" (Eng. trans. G. C. Macaulay—Project Gutenberg, 97).

[19]Diogens Laertius, *Lives*, op. cit., ibidem.

[20]See *Placita Philosophorum* (I, 3) (Eng. trans. William W. Goodwin); see also Diels: *Fragmente*, op. cit., 2.

[21]Aristotle: *On the Heavens* II, 13, 294b (Eng. trans. J. L. Stocks).

[22]It is handed down that he had written a work in prose (lost) entitled *On Nature*.

to the practice of things, but rather to the research. It cannot tell us anything on the (limited) subject we are interested in. In fact, Diogenes Laertius says:

> Regarding the Earth, nevertheless, he does not explain what its characteristics are, and he does not even with regard to the basins. And these are his philosophical opinions.[23]

1.3 Pythagoras and the Pythagorean School

Pythagoras (570–496 BC), born at Samos, was an Ionian and perhaps in his youth had also been a disciple of Anaximander, but this does not imply that one can associate him to the Ionic School. At the age of about forty years he left his native Samos to settle in Croton, in Southern Italy (Magna Graecia), where he founded a school which, actually, had more the features of a political and religious sect than those of a philosophical school. It seems established that he did not leave anything written and all we know on his thought is due to the testimonies (and the imagination!) of the doxographers and of the philosophers of the succeeding centuries, especially of Aristotle. In any case, given the characteristic of his personality, the legends neatly prevail over the historic reality and it is impossible at any rate to separate Pythagoras' thought from the contribution (if it existed) of the components of his School. Obviously, it is not our task, here, to deal with the thought of Pythagoras and the place he takes in the history of science. We only need to introduce the conception of the universe and of the nature of the Earth that Pythagoras had.

Concerning the Pythagoreans, Aristotle said:

> …the so-called Pythagoreans, who were the first to take up mathematics, not only advanced this study, but also having been brought up in it they thought its principles were the principles of all things. Since of these principles numbers are by nature the first, and in numbers they seemed to see many resemblances to the things that exist and come into being more than in fire and earth and water.[24]

Thus, Aristotle holds that the Pythagoreans have attributed to number the function of material cause that the Ionians had attributed to a physical element. The Pythagoreans have attributed to mathematical measure the fundamental function for understanding the order and unity of the world. This undoubtedly contributed to their cosmological conception.

Even if it is not known for certain how Pythagoras reached the conviction that the Earth has a spherical shape (divided in five zones), it is certain that he still considered it to be in the centre of the universe and therefore he cannot be included among the forerunners of Copernicus. To summarize, according to Pythagoras, the universe, the Earth and the other celestial bodies have spherical shapes and the

---

[23]Diogenes Laertius, *Lives*, op. cit., IX, 11.

[24]Aristotle: *Metaphysics* I, 5-985b 23–28 (Eng. trans. W. D. Ross).

Earth is still in the centre. Moreover, the sphere of the fixed stars has a daily rotation from east to west about an axis passing through the centre of the Earth, and the planets have an independent movement of their own in a sense opposite to that of the daily rotation of the fixed stars, i.e., from west to east. All this was held in the time when Pythagoras lived and also immediately after. In fact, according to Aëtius, the opinion of Alcmaeon of Croton, physician and disciple of Pythagoras, was the same: "Alcmaeon and the mathematicians [say] that the planets have a contrary motion to the fixed stars, and in opposition to them are carried from the west to the east".[25]

Things changed with the last Pythagoreans, in particular with Philolaus (470–390 BC) who abandoned the geocentric hypothesis.[26] According to Diogenes Laertius, Philolaus was the first among the Pythagoreans to write a book (*On Nature*).[27] We again resort to Aristotle for the enunciation of the theory:

> Most people—all, in fact, who regard the whole heaven as finite—say it lies at the centre. But the Italian philosophers known as Pythagoreans take the contrary view. At the centre, they say, is fire, and the earth is one of the stars, creating night and day by its circular motion about the centre. They further construct another Earth in opposition to ours to which they give the name counter-earth. In all this they are not seeking for theories and causes to account for observed facts, but rather forcing their observations and trying to accommodate them to certain theories and opinions of their own. But there are many others who would agree that it is wrong to give the Earth the central position, looking for confirmation rather to theory than to the facts of observation. Their view is that the most precious place befits the most precious thing: but fire, they say, is more precious than earth, and the limit than the intermediate, and the circumference and the centre are limits. Reasoning on this basis they take the view that it is not earth that lies at the centre of the sphere, but rather fire. The Pythagoreans have a further reason. They hold that the most important part of the world, which is the centre, should be most strictly guarded, and name it, or rather the fire which occupies that place, the "Guardhouse of Zeus", as if the word "centre" were quite unequivocal, and the centre of the mathematical figure were always the same with that of the thing or the natural centre.[28]

In the cosmological system of Philolaus, the universe is put between Olympus and the immobile central fire (not the Sun). Ten celestial bodies rotate between Olympus and the central fire: the outermost is that which brings the fixed stars, then the five planets known at that time, and at last the Sun, the Moon, the Earth and the counter-earth, which is the nearest one to the central fire and that cannot be seen from the Earth, since the latter is always facing Olympus.

In conclusion, the cosmological system of Philolaus is neither geocentric nor heliocentric.

---

[25]See Aëtius: *Placita Philosophorum* II, 16 (Eng. trans. William W. Goodwin); see also Diels' *Fragmente*, op. cit., 24, 4).

[26]On this see: T. Heath, *Aristarchus of Samos: The Ancient Copernicus* (Oxford, Clarendon Press, 1913 (prt. Dover, 1981) and G. V. Schiaparelli, "I precursori di Copernico nell'antichità" in *Scritti sulla Storia dell'Astronomia antica* (Zanichelli, 1925, rpt. Milan, Mimesis, 1997).

[27]Diogenes Laertius, *Lives*, op. cit., VIII, 85.

[28]Aristotle: *On the Heavens*, II, 293a–293b (Eng. trans. J. L. Stocks).

1.4 The Succeeding Schools, the Atomists, Plato

As we have seen, the new element, consisting in attributing to the Earth a spherical shape, was introduced by Pythagoras and the early Pythagoreans, still hanging on to the geocentric model. Philolaus and the late Pythagoreans (about one century later) will abandon geocentrism.

Obviously, after the Ionic School, about which we have already spoken, and before the late Pythagoreans, in the Greek world several philosophical Schools followed one another: the Eleatic School (Xenophanes, Parmenides, Zeno), the Sophists (Protagoras, Gorgias), those we call the later physicists (Empedocles, Anaxagoras) and finally the Atomists Leucippus and Democritus, contemporaries of Philolaus.

From the point of view of the history of philosophy, we are dealing with fundamental steps of Western thought and, in a few cases, also with advances in the study of the motion of celestial bodies.[29] Nevertheless, there were no new elements regarding the problem we are interested in. For instance, it is not known for certain if Empedocles (who had definitively introduced the four fundamental elements earth, water, air and fire) held the Earth to be spherical or flat. It is more probable that, instead of adopting the view of Pythagorean School and Parmenides, he considered it to be flat, following Anaxagoras.

The theory of spherical Earth was not even accepted by the Atomists. Democritus held it like a disc[30] in equilibrium (since there is no reason why it should move one way rather than another), in contrast to Anaxagoras, who considered it to be supported by the air. As we have seen, this idea has also been attributed to Anaximander, and at last Democritus shared it with Parmenides.[31]

In the generation succeeding to that of the Atomists, we find Plato (427–347 BC). However, as Giorgio de Santillana says,[32] he is too profound a philosopher and too great a writer to analyze in isolation his “contributions to science”, i.e., to extract from the context of his works the fragments of a physical theory and the flashes of not well understood archaic cognitions he scatters here and there in the form of myth. Nevertheless, we cannot neglect our duty of quoting excerpts of his works that allude to the nature and the shape of the Earth. We have chosen two well-known excerpts from *Phaedo* and *Timaeus*.

The first, a part of the dialogue between Simmias and Socrates in the *Phaedo*, says:

---

[29]On this, see T. Heath: *Aristarchus of Samos*, op. cit. and J. L. E. Dreyer *History of the Planetary Systems from Thales to Kepler* (Cambridge University Press, 1906).

[30]Aëtius, *Placita Philosophorum*, III 10, 5 (Diels, op. cit., 68, 94): “Democritus [said] that it is like a quoit in its surface, and hollow in the middle” (Eng. trans. William W. Goodwin).

[31]Aëtius, *Placita Philosophorum*, III 15, 7 (Diels, op. cit. 28, 44): “Parmenides and Democritus, that the earth being so equally poised hath no sufficient cause why it should incline rather to one side than to the other; so that it may be shaken, but cannot be removed.” trans. William W. Goodwin).

[32]G. de Santillana, op. cit, chap. 12.

And there are many marvelous regions of the earth, and she is neither in size nor in any way what she is imagined by those who are used to talking about the earth, as I am persuaded by someone. And Simmias said, "What do you mean by that, Socrates? For I have heard much about the earth myself, yet not the things you believe; so gladly I would listen." "Well then, Simmias, it does not seem to me to be the art of Glaucus to narrate what she is; yet as truth, it appears to me to be harder than by the art of Glaucus, and at the same time perhaps I would not be able to, and besides, even if I knew how, it seems to me my life, Simmias, would be over before the argument is adequate. Yet nothing prevents my telling what I believe to be the form of the earth and her regions." "But," said Simmias, "that will be adequate." "I believe then," he said, "that first, if she is in the middle of the heavens being carried round, she needs neither air nor any other such necessity for her not to fall, but sufficient to maintain her are the likeness of heaven to all of it and the equal balance of the Earth herself; for something equally balanced put in the middle of something similar will not incline at all more nor less, but similarly stays unswerving. First," he said, "I believe this." "And correctly," said Simmias. "Next then," he said, "she is something very large, and those of us living between the pillars of Heracles and the Phasis river live in a small portion around the sea, like ants or frogs around a pond, and many others elsewhere live in many such places. For there are everywhere around the earth many hollows of all sorts both in form and greatness, into which the water and the mists and the air have flowed together; but the earth pure herself is situated in the pure heaven, in which the stars are, which is called ether by many who are used to talking about such things; of which these are the sediment and flow together always into the hollows of the earth. "So our living in her hollows is unnoticed, and we think we are living up on the Earth, just as if someone who lived at the bottom of mid-ocean should think one lived on the sea, and through the water seeing the sun and the other stars should believe the sea to be heaven.[33]

With a prose rich in imagination, the Earth is described as a sphere placed in the centre of the universe and not in need of a support.

The interpretation of the second excerpt, from the *Timaeus*, has caused a heated debate among the scholars. First, let us report the text:

Of the heavenly and divine, he created the greater part out of fire, that they might be the brightest of all things and fairest to behold, and he fashioned them after the likeness of the universe in the figure of a circle, and made them follow the intelligent motion of the supreme, distributing them over the whole circumference of heaven, which was to be a true cosmos or glorious world spangled with them all over. And he gave to each of them two movements: the first, a movement on the same spot after the same manner, whereby they ever continue to think consistently the same thoughts about the same things; the second, a forward movement, in which they are controlled by the revolution of the same and the like; but by the other five motions they were unaffected, in order that each of them might attain the highest perfection. And for this reason the fixed stars were created, to be divine and eternal animals, ever-abiding and revolving after the same manner and on the same spot; and the other stars which reverse their motion and are subject to deviations of this kind, were created in the manner already described. The earth, which is our nurse, clinging around the pole which is extended through the universe, he framed to be the guardian and artificer of night and day, first and eldest of gods that are in the interior of heaven.[34]

---

[33]Plato: *Phaedo* 108–109 (Eng. trans. Sanderson Beck).

[34]Plato: *Timeaus* 39–40 (Eng. trans. B. Jowett).

As J. L. E. Dreyer points out,[35] before arriving at the passage "And he gave to each of them two movements on the same spot ..... the second, a forward movement .... the like", all that is told in the *Timaeus* (as well as in the *Phaedo* and in the *Republic*) leads one to the belief that the fundamental element in Plato's cosmic system is the doctrine of the daily rotation of the heaven about the Earth (in the shape of a sphere, as held by the Pythagoreans) immobile in the centre of the universe. But Aristotle had already remarked: "Others, again, say that the earth, which lies at the centre, is 'rolled', and thus in motion, about the axis of the whole heaven, so it stands written in the *Timaeus*".[36]

The passages from both *On the Heavens* and *Timaeus* have been meticulously analyzed by scholars, as well as, and above all, by philologists, comparing the variances of tradition of the manuscripts and trying to understand if Plato (though in extremis and in an obscure manner) accepted a daily rotation of the Earth or not. The opinion of Dreyer, and also of Duhem,[37] is that Plato considered the Earth as having the shape of a sphere, immobile in the centre of the universe, and supported by nothing less than the symmetry principle.

Of course, we cannot dwell on all the arguments reported in this discussion. For us, it is enough to record that, with Plato, the spherical shape of the Earth begins to be an acquired notion. Obviously, we shall come back, with Aristotle, on this subject widely.

1.5 Aristotle

Before talking about Aristotle and his contributions to the determination of the shape of the Earth, we must briefly mention Eudoxus of Cnidus (408–355 BC), celebrated geometer and astronomer, whose theory of the motions of planets exercised a strong influence on Aristotle, so that he strove to improve it (or, at least, this was his purpose!). The theory of Eudoxus, the first mathematical theory of the motion of planets, is named "of the concentric spheres"; Eudoxus imagined the planets fixed on ideal spheres, uniformly rotating, and all concentric with the Earth. In order to account for the complex motion of planets, he supposed that the sphere of a planet rotated on an axis hinged on another sphere concentric with the former, and so on, by choosing always different axes. By making use of 27 spheres in total he "explained" the motion of the Sun, of the Moon and of the five planets known at that time.[38] As Heath says:

---

[35]J. L. E. Dreyer: *History of the Planetary Systems from Thales to Kepler* (Cambridge University Press, 1906), Chap. 3.

[36]Aristotle, *On the Heavens* II 17, 293b (Eng. trans. J. L. Stocks).

[37]See P. Duhem: *Le Système du Monde* (Paris: Hermann, 1919), vol. 1, chap. II, XI.

[38]Aristotle gives knowledge of this system and of the corrections proposed by Callippus (370–300 BC) in the twelfth book of *Metaphysics* (XII, 8, 1073 b 17–1074 a 14). More information comes from the Commentary of Simplicius (sixth century AD) to *On the Heavens* of Aristotle. On that

.... it was on the theoretic even more than the observational side of astronomy that Eudoxus distinguished himself, and his theory of concentric spheres, by the combined movements of which he explained the motions of the planets (thereby giving his solution of the problem of accounting for those motions by the simplest of regular movements), may be said to be the beginning of scientific astronomy.[39]

Eudoxus' system was merely geometric and theoretic. Instead, Aristotle (384–322 BC), who (differently from Plato) investigated the "idea" in its concrete realization in the phenomena of nature, transformed the abstract and geometric theory of Eudoxus into a mechanical system of spheres, i.e., of spherical shells, in real contact one with another. This fact practically obliged him to augment the number of spheres of Callippus and then to construct a quite complicated mechanical system.

Obviously, to be concerned with the system of the concentric spheres lies outside our intent, but it is inside this system that Aristotle's theory of Earth is inserted, also including the conclusions on its spherical shape.

Let us see how Aristotle tackles the problem, by directly resorting to his writings. As we have already recalled, *Metaphysics* is the work in which Aristotle expounds the system of Eudoxus and his disciple Callippus, but it is in *On the Heavens* (Book II) that he widely treats the sphericity of the universe and, as a consequence of the Earth. Let us begin from the heaven:

The shape of the heaven is of necessity spherical; for that is the shape most appropriate to its substance and also by nature primary. First, let us consider generally which shape is primary among planes and solids alike. Every plane figure must be either rectilinear or curvilinear. Now the rectilinear is bounded by more than one line, the curvilinear by one only. But since in any kind the one is naturally prior to the many and the simple to the complex, the circle will be the first of plane figures. Again, if by complete, as previously defined, we mean a thing outside which no part of itself can be found, and if addition is always possible to the straight line but never to the circular, clearly the line which embraces the circle is complete. If then the complete is prior to the incomplete, it follows on this ground abo that the circle is primary among figures. And the sphere holds the same position among solids.[40]

We have quoted this excerpt in order to show how deeply the Platonic "religion" of circularity permeated the work of Aristotle and, as we shall see, that of his successors. It will take another nineteen centuries to break the circle (with Kepler, in 1609).[41]

---

report Schiaparelli based his reconstruction of the "mathematical mechanism" imagined by Eudoxus. See G. V. Schiaparelli: "Le Sfere omocentriche di Eudosso, di Callippo e di Aristotele" in *Scritti sulla storia dell'astronomia antica*, op. cit., II, pp 5–112.

[39]See T. L. Heath: *Aristarchus of Samos*, op. cit., p. 193.

[40]Aristotle: *On the Heavens*, op. cit., II, 4, 286 b (Eng. trans. J. L. Stocks).

[41]See D. Boccaletti: "From the epicycles of the Greeks to Kepler's ellipse. The breakdown of the circle paradigm in Cosmology through time", in *Cosmology Through Time: Ancient and Modern Cosmology in the Mediterranean Area* (Conference proceedings, Monte Porzio Catone (Rome), Italy, June 18–20, 2001), eds. S. Colafrancesco and G. Giobbi (Milan: Mimesis) pp 99–112.

Further on, he expounds the mechanism of the concentric spheres. We limit ourselves to an indicative passage:

> Now the first figure belongs to the first body, and the first body is that at the farthest circumference. It follows that the body which revolves with a circular movement must be spherical. The same then will be true of the body continuous with it: for that which is continuous with the spherical is spherical. The same again holds of the bodies between these and the centre. Bodies which are bounded by the spherical and in contact with it must be, as wholes, spherical; and the bodies below the sphere of the planets are contiguous with the sphere above them. The sphere then will be spherical throughout; for every body within it is contiguous and continuous with spheres.[42]

The sphericity of the universe also requires that the surface of the water on the Earth be a spherical surface:

> Therefore, if the heaven moves in a circle and moves more swiftly than anything else, it must necessarily be spherical. Corroborative evidence may be drawn from the bodies whose position is about the centre. If earth is enclosed by water, water by air, air by fire, and these similarly by the upper bodies—which while not continuous are yet contiguous with them—and if the surface of water is spherical, and that which is continuous with or embraces the spherical must itself be spherical, then on these grounds also it is clear that the heavens are spherical. But the surface of water is seen to be spherical if we take as our starting-point the fact that water naturally tends to collect in a hollow place—"hollow" meaning "nearer the centre". Draw from the centre the lines AB, AC, and let their extremities be joined by the straight line BC. The line AD, drawn to the base of the triangle, will be shorter than either of the radii. Therefore the place in which it terminates will be a hollow place. The water then will collect there until equality is established, that is until the line AE is equal to the two radii. Thus water forces its way to the ends of the radii, and there only will it rest: but the line which connects the extremities of the radii is circular: therefore the surface of the water BEC is spherical.[43] (see Fig. 1.1).

We go on now to look at the point where Aristotle specifically speaks about the Earth (quoting again part of the excerpt from *On the Heavens* we have already seen):

> It remains to speak of the earth, of its position, of the question whether it is at rest or in motion, and of its shape. As to its position there is some difference of opinion. Most people—all, in fact, who regard the whole heaven as finite—say it lies at the centre. But the Italian philosophers known as Pythagoreans take the contrary view. At the centre, they say, is fire, and the earth is one of the stars, creating night and day by its circular motion about the centre. They further construct another Earth in opposition to ours to which they give the name counterearth. In all this they are not seeking for theories and causes to account for observed facts, but rather forcing their observations and trying to accommodate them to certain theories and opinions of their own. But there are many others who would agree that it is wrong to give the earth the central position, looking for confirmation rather to theory than to the facts of observation.[44]

---

[42]Aristotle: *On the Heavens* op. cit., II, 4, 287 a (Eng. trans. J. L. Stocks).

[43]Aristotle: *On the Heavens* II, 287 a–287 b (Eng. trans. J. L. Stocks).

[44]Aristotle: *On the Heavens* II, 13, 293 a.

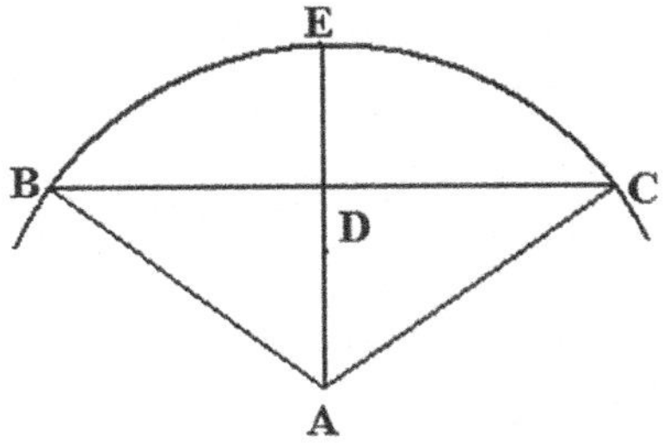

Fig. 1.1 The spherical surface of water

Of course, he too does not realize that he himself, by following the "religion of the circle", makes assertions not based on phenomena. Thus, on the immobility of the Earth:

> Let us first decide the question whether the earth moves or is at rest. For, as we said, there are some who make it one of the stars, and others who, setting it at the centre, suppose it to be "rolled" and in motion about the pole as axis. That both views are untenable will be clear if we take as our starting-point the fact that the earth's motion, whether the earth be at the centre or away from it, must needs be a constrained motion. It cannot be the movement of the earth itself. If it were, any portion of it would have this movement; but in fact every part moves in a straight line to the centre. Being, then, constrained and unnatural, the movement could not be eternal. But the order of the universe is eternal. Again, everything that moves with the circular movement, except the first sphere, is observed to be passed, and to move with more than one motion. The earth, then, also, whether it move about the centre or as stationary at it, must necessarily move with two motions. But if this were so, there would have to be passings and turnings of the fixed stars. Yet no such thing is observed. The same stars always rise and set in the same parts of the earth.
>
> ...
>
> Further, the natural movement of the earth, part and whole alike, is the centre of the whole —whence the fact that it is now actually situated at the centre—but it might be questioned since both centres are the same, which centre it is that portions of earth and other heavy things move to. Is this their goal because it is the centre of the earth or because it is the centre of the whole? The goal, surely, must be the centre of the whole. For fire and other light things move to the extremity of the area which contains the centre. It happens, however, that the centre of the earth and of the whole is the same. Thus they do move to the centre of the earth, but accidentally, in virtue of the fact that the earth's centre lies at the centre of the whole. That the centre of the earth is the goal of their movement is indicated by the fact that heavy bodies moving towards the earth do not parallel but so as to make equal angles, and thus to a single centre, that of the earth. It is clear, then, that the earth must be at the centre and immovable, not only for the reasons already given, but also because heavy bodies forcibly thrown quite straight upward return to the point from which they started, even if they are thrown to an infinite distance. From these considerations then it is clear that the earth does not move and does not lie elsewhere than at the centre.[45]

[45]Aristotle: *On the Heavens* II, 14, 296 a–296 b (Eng. trans. J. L. Stocks).

This excerpt also contains the statement that the bodies falling on the Earth do not move on parallel straight lines but on lines all converging at the centre (of the Earth). In fact, the expression "equal angles", as unanimously acknowledged by the scholars, must be interpreted as equal (right) angles with the tangents to the (spherical) Earth. Obviously, it remains to question on what experiments Aristotle based this (correct) conclusion. Let us arrive, finally, at the part in which we are more interested, the shape of the Earth:

> Its shape must necessarily be spherical. For every portion of earth has weight until it reaches the centre, and the jostling of parts greater and smaller would bring about not a waved surface, but rather compression and convergence of part and part until the centre is reached. The process should be conceived by supposing the earth to come into being in the way that some of the natural philosophers describe. Only they attribute the downward movement to constraint, and it is better to keep to the truth and say that the reason of this motion is that a thing which possesses weight is naturally endowed with a centripetal movement. When the mixture, then, was merely potential, the things that were separated off moved similarly from every side towards the centre. Whether the parts which came together at the centre were distributed at the extremities evenly, or in some other way, makes no difference. If, on the one hand, there were a similar movement from each quarter of the extremity to the single centre, it is obvious that the resulting mass would be similar on every side. For if an equal amount is added on every side the extremity of the mass will be everywhere equidistant from its centre, i.e. the figure will be spherical. But neither will it in any way affect the argument if there is not a similar accession of concurrent fragments from every side. For the greater quantity, finding a lesser in front of it, must necessarily drive it on, both having an impulse whose goal is the centre, and the greater weight driving the lesser forward till this goal is reached.[46]

This is substantially the first proof Aristotle proposes for demonstrating the sphericity of the Earth. P. Duhem remarks that, even if in a vague form, here it is asserted that it is gravity that determines the sphericity: "*C'est à la pesanteur que la terre doit sa figure*".[47]

> But the spherical shape, necessitated by this argument, follows also from the fact that the motions of heavy bodies always make equal angles, and are not parallel. This would be the natural form of movement towards what is naturally spherical. Either then the earth is spherical or it is at least naturally spherical. And it is right to call anything that which nature intends it to be, and which belongs to it, rather than that which it is by constraint and contrary to nature. The evidence of the senses further corroborates this. How else would eclipses of the moon show segments shaped as we see them? As it is, the shapes which the moon itself each month shows are of every kind straight, gibbous, and concave—but in eclipses the outline is always curved: and, since it is the interposition of the earth that makes the eclipse, the form of this line will be caused by the form of the earth's surface, which is therefore spherical.[48]

---

[46]Aristotle: *On the Heavens* II, 14, 297 a (Eng. trans. J. L. Stocks).

[47]P. Duhem: *Le système du monde*, op. cit., p. 213.

[48]Aristotle: *On the Heavens* II, 14, 297 b (Eng. trans. J. L. Stocks).

The last argument, which undoubtedly is, as nowadays we should say, the "most scientific", was not appreciated in Antiquity. In fact, the historian Paul Tannery points out:

Il est particulièrement singulier que l'on ne retrouve ni dans Ptolémée ni dans les cosmographes élémentaires l'argument le plus sérieux que l'antiquité ait connu, argument que pourtant Aristote avait déjà mis en avant, à savoir que, dans les éclipses de lune, la limite de l'ombre de la terre affecte toujours la forme circulaire. Sans doute cette preuve avait été écartée comme pouvant difficilement être invoquée au début d'une exposition méthodique, peut-être comme exigeant, pour son développement, un appareil géométrique incompatible avec les éléments de la science.[49]

At last, we report the excerpt regarding the size of the Earth:

Again, our observations of the stars make it evident, not only that the earth is circular, but also that it is a circle of no great size. For quite a small change of position to south or north causes a manifest alteration of the horizon. There is much change, I mean, in the stars which are overhead, and the stars seen are different, as one moves northward or southward. Indeed there are some stars seen in Egypt and in the neighbourhood of Cyprus which are not seen in the northerly regions; and stars, which in the north are never beyond the range of observation, in those regions rise and set. All of which goes to show not only that the earth is circular in shape, but also that it is a sphere of no great size: for otherwise the effect of so slight a change of place would not be quickly apparent. Hence one should not be too sure of the incredibility of the view of those who conceive that there is continuity between the parts about the pillars of Hercules and the parts about India, and that in this way the ocean is one. As further evidence in favour of this they quote the case of elephants, a species occurring in each of these extreme regions, suggesting that the common characteristic of these extremes is explained by their continuity. Also, those mathematicians who try to calculate the size of the earth's circumference arrive at the Fig. 400,000 stades. This indicates not only that the earth's mass is spherical in shape, but also that as compared with the stars it is not of great size.[50]

With regard to the estimate of 400,000 stades (which, in the hypothesis that a stade is equivalent to 183 m, would amount to 73,000 km) for the circumference of the Earth, scholars assume that Aristotle reported the result of a calculation in all probability due to Eudoxus. Obviously, 73,000 km is almost twice the actual value of the Earth's circumference.

As we shall see later, it will fall to Eratosthenes to get nearer to the real value. As the reader can undoubtedly have remarked, with Aristotle we have entered into a scheme of things completely different from that of his predecessors. As regards his conclusion on the shape of the Earth, Aristotle puts forward proofs, such as the one based on the lunar eclipses, that come from a deductive reasoning.

We have no more to do with the intuitions of the Pythagoreans. We can conclude that, from Aristotle on, the sphericity of the Earth is an acquired and also practically shared concept, at least in the ambit of natural philosophers.

[49]See Paul Tannery: *Recherches sur l'histoire de l'astronomie ancienne* (Paris, 1893), p. 103.

[50]Aristotle: *On the Heavens* II, 14, 297b–298a (Eng. trans. J. L. Stocks).

1.6 Pytheas' Travels and Early Greek Geography

Until now, we have dealt essentially with the "theoretical" elaboration regarding the shape of the Earth drawn up by the several philosophers who came one after the other from the seventh to the fourth centuries BC. Even if, with Aristotle, some "demonstrations" were put forward, however there was some distance between real direct observations and concrete evidence.

When talking about Anaximander we hinted at the fact that he was credited with the invention of the gnomon and of the calculation of the equinoxes, but we did not enter into a detailed discussion of either. We shall do this now, introducing the narrations of the travels and of the explorations which were made, especially in the fourth century, and mainly tied to the name of Pytheas of Massalia. First of all, we must say that the action of associating the flow of time in the course of the day with the variation of the length of the shadow of an object fixed in the ground is a practice which goes back to the deep recesses of time. Thus, it is impossible to say when the use of the gnomon began. We can also say that the habit of observing the lengthening of shadows for evidence of the approaching of evening lasted as long as humankind had no mechanical clocks. Even as late as Augustan Rome, Virgil ended his first Eclogue with the marvellous distich:

... et iam summa procul villarum culmina fumant

maioresque cadunt altis de montibus umbrae.[51]

At this point, for introducing our subject, it is worthwhile to go on with the discussion starting again from Anaximander. Whereas, from what is reported by several authors, it is sufficiently attested that Thales, presumably making use of the gnomon, already knew the position of the solstices on the meridian line, it is uncertain what Anaximander's knowledge on the equinoxes was.

On this question, which in some way has to do with establishing a (historical!) explanation of the fact that Pytheas (about the thirties of the fourth century) could have had knowledge, both theoretical and practical, of the use of the gnomon for determining the latitude of the different places, Á. Szabó[52] puts forward a hypothesis made up of a series of conjectures strongly supported by their internal logic and based on a reasonable interpretation of the documents. He points out that subsequent to the information supplied by Diogenes Laertius, there were the attestations by the bishop of Cesarea Eusebius (sixth century) in his work *Praeparatio Evangelica* as well as, and more importantly, the Byzantine Greek

[51]"And, see, the farm-roof chimneys smoke afar/ And from the hills the shadows lengthening fall!" (Eng. trans. John William Mackail).

[52]Á. Szabó and E. Maula: *Le début de l'astronomie de la géographie et de la trigonométrie chez les Grecs* (Paris: Vrin, 1986), première partie.

Suda encyclopaedia (tenth century), both of which are fairly explicit on the question of equinoxes.[53] Before going further let us try to expose the terms of the question.

By making use of the gnomon, a pole of a determined length fixed perpendicularly in the ground (i.e., directed towards the centre of the Earth), one can determine the extreme of the shadow when the sun is in the zenith in the various seasons of the year. Since the segment of shadow rotates during the day, one fixes the direction given by the meridian shadow when it reaches the smallest distance during the year. In this way one obtains the position of the summer solstice. On the same meridian line, which represents the north-south direction, and therefore the terrestrial meridian through that place, one can also determine the point where the shadow reaches the largest length, i.e. the position of the vernal solstice. This entire procedure requires only an observation (at noon when the sun is in the zenith) every day for the course of a full year. In this way, one determines the so-called inter-solstitial arc, or, better, its projection on the Earth. From the terrestrial segment of the meridian line one can easily obtain the chord of this arc.[54]

Now, the problem for determining the equinoxes and the obliquity of the ecliptic is that of the bisection of the chord obtained. For doing this, knowledge, even if elementary, of trigonometry is necessary. We recall that the Greeks, for the problems concerning the triangle, did not use our usual trigonometric functions but the chord of the arc (of the unit circle) having as a central angle the angle they wanted to calculate (the relation with the modern trigonometric functions is chord $\alpha = 2\sin\frac{\alpha}{2}$, where chord α indicates the length of the segment representing the chord). It is this knowledge that Szabó ascribes to Anaximander, despite the fact that no account at all has been left of how he succeeded in obtaining the determination of the equinoxes credited to him.

Szabó's conclusions are these: Pytheas could not carry out his travels and perform his measurements (latitude of the places he visited, etc.) without having as a prerequisite the necessary astronomical and mathematical knowledge. No testimonies are left which attest the development of the knowledge on the use of the gnomon and the determination of the equinoxes relevant to the three centuries that elapsed between Anaximander and Pytheas. Indeed, it seems that these things are taken for granted, as if there were no need of referring them to somebody or to some School. Therefore, Szabó says, this knowledge (by interpreting Diogenes Laertius, Eusebius and Suda) must be credited to Anaximander, and it must be considered that its later use caused it to become a common heritage for astronomers and mathematicians.

---

[53]Here is the text: "Son of Praxiades, Milesian, philosopher, a relative and student and successor of Thales. He first discovered an equinox and solstices and hour-indicators, and that the Earth is situated in the middle [of the universe]. He also introduced a sundial and explained the basis of all geometry. He wrote *On Nature*, *Circuit of the Earth*, and *Fixed Bodies* and *Globe* and some other works" (Eng. trans. Jennifer Benedict, From Suda on Line: Byzantine Lexicography—http://www.stoa.org/sol).

[54]We refer the reader to Szabó and Maula, *Le début de l'astronomie de la géographie*, op. cit., for all the geometric treatment.

And now, let us come finally to Pytheas.[55] As we have said above, he was a Greek, but lived in the Phocaean colony of Marseille (Massalia), one of the first colonies established by the Greeks in the Mediterranean Sea. His biographical data are exceptionally scanty: even the date of his famous voyage is debated among scholars, as is the exact period in which he lived. As far as we are concerned, we consider have period quoted above to be the most probable one. One must remember that this is also the period of the conquests of Alexander the Great (356–329 BC) and that Alexander sponsored the institution of fleets for the exploration of lands not yet conquered.[56]

Most of the scholars consider it to be almost verified that Pytheas knew Eudoxus' theory of spheres; indeed, some go as far as to presume that Pytheas was a pupil of Eudoxus. Besides knowing Eudoxus' model of the universe, which assumed a spherical Earth, Pytheas (as we have already recalled) was extraordinarily skilful in the use of the gnomon. During his travels to the extreme North, with his observations, he found confirmation of Eudoxus' theory, investigated the length of the day, the height of the sun in the solstices, and the position of the celestial pole of the northern hemisphere, through which the axis of the cosmos (and of the Earth in its centre) passed. He determined the latitude of several places and studied the tides. His voyage had as a first destination Càdiz (in the neighbourhood of the Pillars of Heracles) and as a final destination the extreme North, following the Atlantic coast of Europe, as far as the British Islands and at last reaching Thule (an island never really identified and which then became a mythic place for indicating the boundary of the oecumene; only recently has it been demonstrated that Thule is very probably the Norwegian island Smola[57]).

He recounted all this in a work, now lost and of which only fragments are extant (quoted by other authors), entitled *On the Ocean*. Pytheas' book found a large readership both among his contemporaries and the geographers of succeeding centuries. Pytheas' story was not always believed; on the contrary, the most famous geographer of Antiquity, Strabo (64 BC–25 AD), about three centuries later, challenged much of the results reported by Pytheas. Instead, in the modern age, he won credibility. In any case, he had an important appreciator in Eratosthenes (276–194 BC), who is considered the founder of the scientific geography.

We have given room to Pytheas' travels, even if it is not our purpose to deal with the history of geography, since his work can be counted among the most important evidence for the belief in the sphericity of the Earth at that time.

---

[55]On Pytheas and his travels we refer the reader to the thorough book (in Italian) of Stefano Magnani, *Il viaggio di Pitea sull'oceano* (Bologna: Patron editore, 2002).

[56]In the fourth century many voyages (even with non-military purposes) were performed by Greek navigators, but that of Pytheas is the most important both for the lands he visited and the detailed account he drew up of it.

[57]See Andreas Kleineberg, Christian Marx, Eberhard Knobloch, Dieter Leigemann, *Germania und die Insel Thule* (Darmstadt, 2010), Chap. 6.

1.7 Eratosthenes' Measurement of the Earth

Before talking about the work of Eratosthenes regarding the shape of the Earth and the measurement of its size, we cannot avoid quoting the work of Aristarchus of Samos (about 310–230 BC). Preceded by Heraclides of Pontus (385–322 or 310 BC), who had already put forward a partial heliocentric hypothesis that also attributed to the Earth a diurnal rotation motion, Aristarchus is known as the brave defender of the heliocentric hypothesis in Antiquity; in fact, T. L. Heath called him "the ancient Copernicus". He also dealt, in particular, with evaluating the distances of the Sun and of the Moon (to us his work remains with the title *On the sizes and distances of the sun and moon*[58]).

Archimedes (287–212 BC), about three decades his junior, recalls his "Copernican" system in his famous work *The Sand Reckoner*:

> ... But Aristarchus of Samos brought out a book consisting of some hypotheses, in which the premises lead to the result that the universe is many times greater than that now so called. His hypotheses are that the fixed stars and the sun remain unmoved, that the earth revolves about the sun in the circumference of a circle, the sun lying in the middle of the orbit.[59]

A further unequivocal evidence is a passage of Plutarch:

> Only do not, my good fellow, enter an action against me for impiety in the style of Cleanthes, who thought it was the duty of Greeks to indict Aristarchus of Samos on the charge of impiety for putting in motion the earth of the Universe, this being the effect of his attempt to save the phenomena by supposing the heaven to remain at rest and the earth to revolve in an oblique circle, while it rotates, at the same time, about its own axis.[60]

Although serious authors have shown that Ptolemy used Aristarchus' system in order to elaborate his theory, of course without quoting its author, Aristarchus, the theory had no effect in the centuries to come and his heliocentric model was abandoned. We limited ourselves to a few mentions, since this is not the subject of the present book. Instead, it is important for us to dwell on an invention, or better, the discovery of an improvement of the gnomon, as reported by Vitruvius.[61] This is the so-called scaphe (σκὰφη) which was, not a plane, but a concave hemispherical surface, with a pole erected vertically in the middle (see Fig. 1.2). With this equipment and by means of lines marked on the surface of the hemisphere one was able to read the direction and the height of the sun with sufficient accuracy. Eratosthenes made good use of this instrument, as we shall try to narrate.

---

[58]See the English translation in T. L. Heath's *Aristarchus of Samos*, op.cit., ed. 1981, pp. 353–414.

[59]T. L. Heath (ed.): *The Work of Archimedes* (Cambridge University Press, 1897; rpt. Dover, 2002), pp. 221–222.

[60]Plutarch: *De facie in orbe lunae,* op. cit., p. 304 (see the translation by T. L. Heath in *Aristarchus of Samos*, op. cit.).

[61]Vitruvius: *De architectura* IX, 8, 1. According to Szabò, "*Vitruve s'est trompé*" (Szabò and Maula, *Le début*, op. cit., p. 60). On the basis of quotations from Aristophanes, he holds that the scaphe existed before Aristarchus.

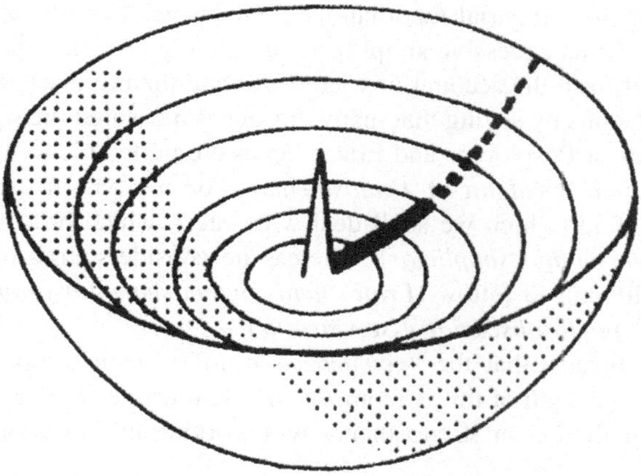

Fig. 1.2 The scaphe

We have only scanty biographical information of Eratosthenes of Cyrene, supplied by Cleomedes, Pliny, Strabo, Ptolemy, and others among the writers of Antiquity. He was born in Cyrene in 276 BC and died in Alexandria in 194 BC. After a philosophical education in Athens, when he was about thirty years old he was invited to Alexandria by King Ptolemy III (Evergetes I), possibly at the instigation of his fellow countryman Callimachus, who had already been given a post in the library by King Ptolemy II (Philadelphus). On the death of the first chief librarian Zenodotus, ca. 235, Eratosthenes was appointed to the post, where he remained until his death.

In his work he touched various fields (mathematics above all) and wrote on many subjects, but only fragments of his writings are extant and what is known has been reconstructed from the memoirs and the quotations of his contemporaries and of the philosophers of the succeeding centuries. Since we are interested only in the work regarding what nowadays would be called geodetic surveys,[62] our primary source will be the work of Cleomedes,[63] even though Cleomedes' narration of the

[62]Regarding this, J. Dutka ("Eratosthenes' measurement of the Earth reconsidered", *Arch. Hist. Exact Sci.* 46 (1993): 55–66) points out that the method used by Eratosthenes, i.e. the use of the proportion

$$\frac{\textit{circumference of the Earth}}{\textit{terrestrial distance}} = \frac{\textit{angular measure of the meridian}}{\textit{angular difference of latitude}},$$

has remained the basis procedure for geodetic surveys, except some technical improvements notably in the seventeenth century, until the advent of geodesic satellites (ca. 1960).

[63]Little is known of the life of Cleomedes. He seems to have written his elementary astronomical textbook *De motu circulari corporum caelestium* between the time of Posidonius and Ptolemy (i.e., between the first century BC and the second century AD). The book was written in Greek; we have quoted the Latin title of the classical translation of H. Ziegler (Teubner, Leipzig, 1891). A partial English translation can be found in Heath's *Greek Astronomy* (London, 1932; rpt. Dover,

measurement of the terrestrial meridian by Eratosthenes has been criticised, especially recently, for its excessive simplifications. In any case, it is more convenient for us to first set forth the account of Cleomedes and then to discuss the criticisms.

Cleomedes begins by saying that many physicists measured the size of the Earth, but the measures of Posidonius and Eratosthenes are preferable to the rest (*reliquis probabiliores sunt Posidonii et Eratosthenis*). For his part he seems to prefer Posidonius' method (which we shall deal with later), which he considers simpler (*Posidonii autem opinio simplicior*), whereas he considers Eratosthenes' method slightly more difficult to follow (*Eratosthenis autem sententia geometricam rationem sequitur et paulo obscurior esse videatur*).

It is a little bizarre that the "*geometrica ratio*" is considered to be obscure! Anyway, on the strength of this, Cleomedes first enumerates five points for making Eratosthenes' method clear (Eratosthenes was working in Alexandria):

> Let us suppose, in this case too, first, that Syene and Alexandria lie under the same meridian circle; secondly, that the distance between the two cities is 5,000 stades; and thirdly, that the rays sent down from different parts of the sun on different parts of the earth are parallel; for this is the hypo-thesis on which geometers proceed. Fourthly, let us assume that, as proved by the geometers, straight lines falling on parallel straight lines make the alternate angles equal, and fifthly, that the arcs standing on (i.e., subtended by) equal angles are similar, that is, have the same pro-portion and the same ratio to their proper circles—this, too, being a fact proved by the geometers. Whenever, therefore, arcs of circles stand on equal angles, if any one of these is (say) one-tenth of its proper circle, all the other arcs will be tenth parts of their proper circles.[64]

After these preliminaries, Cleomedes dwells on explaining to his readers, presumed not to be expert in Euclidean geometry, some elementary results to substantiate Eratosthenes' method. If one accepts that Alexandria and Syene (nearby to modern Aswan) are on the same meridian and the distance Alexandria-Syene is known and, further, that on the day of the summer solstice at Syene the gnomon has no shadow, then the situation is that depicted in Fig. 1.3.

In the figure, the circle with centre O represents the meridian through Alexandria (A) and Syene (S), arc AS the distance between Alexandria and Syene (at that time it was known and measured), angle θ the angle between the direction of the sunbeams and the vertical as inferred from the length of the shadow of the gnomon at Alexandria at noon in the summer solstice.

If Eratosthenes had used a common gnomon, angle θ should have been calculated through the relation $\theta = \arctan\frac{s}{h}$ (see Fig. 1.4), where h is the height of the gnomon and s the length of the shadow.

1981). A recent French translation exists: Cléomède, *Théorie Élementaire*, trans. and comm. R. Goulét (Paris: Vrin, 1980). (We owe to this book Fig. 1.2.) We must consider that Cleomedes' work was a popular work and that at that time Eratosthenes' works *Geography* and *The Measurement of the Earth* were certainly still accessible to the scholars. Unfortunately, Cleomedes' book is the only complete account of Eratosthenes' measurements still extant, thus we must start from it.

[64]The translation of this passage is from T. L. Heath: *Greek Astronomy*, op. cit., pp 109–110.

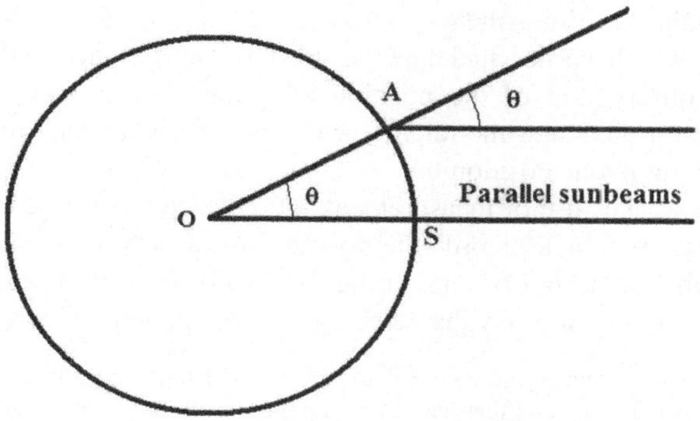

Fig. 1.3 The method of Eratosthenes for determining the circumference of the Earth

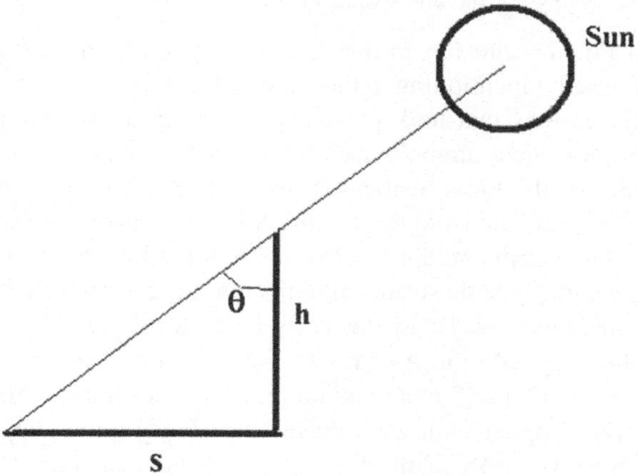

Fig. 1.4 Calculation of angle θ by a gnomon

The use of the scaphe enabled Eratosthenes (since the ancient Greeks did not have trigonometry as we know today) to read the angle directly. Once one has determined what fraction the angle θ is of 360°, one also obtains what fraction of the meridian the arc AS is and can immediately calculate the length of the meridian. According to the narration of Cleomedes, Eratosthenes has obtained for angle θ one-fiftieth of 360° (corresponding to 7° 12′) and accepted for distance AS (Alexandria-Syene) 5,000 stades. As a consequence, the complete great circle (i.e., the terrestrial meridian) measures $50 \times 5,000 = 250,000$ stades.

As we have already said, Cleomedes' piece of writing is the only complete account, even if extremely succinct and simplified, still extant regarding the measurement of the terrestrial meridian performed by Eratosthenes. We also have other information quoted by several authors, even in the Middle Ages, but it is often fragmentary and not as extensive as the narration of Cleomedes. As we shall see later on, he also reported the measurements performed by Eratosthenes during the winter solstice (again in the two towns Syene and Alexandria).

The originality of Eratosthenes' method was in transferring onto the Earth measurements which earlier had been instead of an astronomical nature, being based on the observation of the position of given stars in the sky over towns assumed to lie on the same meridian[65] (we shall deal with measurements of this kind when talking about Posidonius).

Eratosthenes makes use of measurements of distances performed on the ground, and of the gnomon, which records the position of the sun in the sky over a given place at a given instant, but by measuring the shadow thrown by the gnomon and thus not observing the sky. As the celebrated mathematician George Pólya says:

> That Eratosthenes' result is inaccurate does not really detract from the greatness of his achievement. It is his method that excites our admiration. Would not a giant measure the Earth by encircling it with his arms to compare its circumference with his span? And what did our little pygmy Eratosthenes do? At Alexandria at noon on a certain midsummer's day long ago, he observed the shadow cast by a little stick and used his protractor. A mere shadow and the pygmy a giant who spanned the Earth.[66]

Referring to Pólya's allusion to the scarce precision of the result, we take the opportunity of briefly mentioning (since the subject is a matter for specialized scholars) the discussions that took place about the figures reported above. Let us begin from the unit of measurement used for the distances, the stade. As everybody knows, the origin of the units of measurement of length is of anthropomorphous nature (foot, cubit, pace, etc., and their multiples and divisions) and at any time and in any place the same term could indicate units of different length. At present, when this is possible, we translate these ancient units into metres and relative multiples. In the case of the stade used by Eratosthenes, a debate was open (and is still open) as to the value to be assigned to it. As a result, the value given for the stade can range from a minimum of 157.5 m[67] to a maximum of 185 m (Attic stade). This last value is actually the very popular, but each thesis finds fairly good support in the quotations of ancient sources. Recently, Lucio Russo[68] has obtained a confirmation of the value suggested by Friedrich Hultsch. He started from the correction of errors present in more than 6000 longitudes calculated by Ptolemy in his *Geographia*. His analysis regarded a sample of towns for which the ancient data of the coordinates were reliable enough. From the analysis of the distortion of longitudes he succeeded in going back to Eratosthenes' stade, finding a value of 155.6 m.

---

[65] A measurement of this type, reported by Cleomedes as anonimous, is ascribed to Dicearchus of Messina (350–290 BC) and considers lying on the same meridian the two towns of Syene and Lysimachia (Gallipoli on the Dardanelles). Maybe Archimedes refers to this measurement when (in *The sand-reckoner*) he alludes to a few who tried to demonstrate that the perimeter of the Earth measures 300,000 stades.

[66] G. Pólya: *Mathematical Methods in Science* (Washington, 1977).

[67] See F. Hultsch: *Griechische und römische Metrologie* (Berlin: Weidman, 1882).

[68] L. Russo: "Ptolemy's Longitudes and Eratosthenes' measurement of the Earth circumference", *Mathematics and Mechanics of complex Systems* 1(2013): 67–79.

Furthermore, whereas Cleomedes reports the value of 250,000 stades, in the literature subsequent to Eratosthenes the value of 252,000 stades is reported (obviously attributing it to Eratosthenes). One supposes that Cleomedes, for the sake of simplicity, has rounded the figures. In conclusion, according to the value one attributes to the stade used by Eratosthenes in the measurement of the meridian, one goes from an error (with respect to the actual value of the length of the meridian) of little more than 1% to one of 17%. It is obvious (but it must be pointed out) that the measurement of Eratosthenes assumes a perfectly spherical Earth.

As we have recalled above, Cleomedes also tells of measurements made by Eratosthenes[69] during the winter solstice, again in Alexandria and Syene. In this case, since not even Alexandria lies on the winter tropic (*propterea quod haec urbs a tropico hiberno longius distat*), the measurement consisted in comparing (in the same day and at the same time) the length of the shadows in the two towns. The difference still resulted in 1/50 of 360° (*quinquagesima pars maximi circulorum orologii*). This circumstance testifies that Eratosthenes' measurements were part of a wide-ranging project of geodetic surveys, perhaps supported by the government of the Egypt.

At the end of chapter X of his book, Cleomedes makes this assertion: *Erit igitur terrae diametrus supra octoginta milia, quippe quae tertiam partem maximi circuli eius contineat*; i.e., in plain words, the diameter is a third of the circumference. This, as Lucio Russo points out,[70] must induce us to think that the figures given by Cleomedes are rough approximations in order to give an easily understandable explanation.

Coming to a close and also considering the appreciations of the ancient (and medieval) writers regarding the measurements of Eratosthenes (except for Strabo, about whom we shall speak later on), it is unquestionable that we are in presence of a result that nowadays would be called "scientific". In fact, as Giorgio Dragoni remarks, the method used by Eratosthenes possesses, even if in embryo, some of the features of the modern scientific method, such as the repeatability of the experiment. Another important element, according to Dragoni, is that in Eratosthenes' measurements the expertise of the Egyptian geodesists, specialized in the measurement of the terrestrial distances for re-establishing the borders erased by the flooding of the Nile, were employed. Thus there was a synergy of the mathematics (by Eratosthenes) with experience.

---

[69]G. Dragoni, in his book *Eratostene e l'apogeo della scienza greca* (Bologna: CLUEB, 1979), which is one of the more thorough works about Eratosthenes, holds with sound arguments that the winter measurements of which Cleomedes does not mention the author must be credited to Eratosthenes. Dragoni rightly points out that most of the scholars neglect this part of the narration of Cleomedes. Maybe, one can suppose, that many, rather than to go back to the original work, are content with referring to quotations from others.

[70]L. Russo, loc. cit.

1.8 Graeco-Roman Civilization and the Decline of Science

Eratosthenes was one of the peaks of excellence of the Hellenistic science, which obtained its maximum results just straddling the third century BC. Afterwards, from the beginning of the second century, what one can call the Graeco-Roman civilization began. The Romans, victorious in the political and military ambits, approached Greek culture and progressively absorbed it (as Horace wrote, "*Graecia capta ferum victorem cepit*"[71]). But this process only concerned what we would call nowadays humanistic culture, whereas scientific culture and research underwent a decline.[72]

In the Hellenistic, period, what achieved a great success was given by the popularization of the scientific results (and took the place of the scientific treatises) due to authors who, in their compilations, often debated topics about which they were not qualified to speak. For the Romans, this was sufficient, as they probably did not consider scientific research to be necessary, and making the best of the results which could have an immediate practical application.[73] Who best expresses this approach is Cicero:

> *in summo apud illos honore geometria fuit, itaque nihil mathematicis inlustrius; at nos metiendi ratiocinandique utilitate huius artis terminavimus modum.*[74]

Over half a century before Cicero we meet with a figure who well interpreted the spirit of the age, Crates of Mallos (flourished around 160 BC). Crates was the chief librarian of the library of Pergamum and one of the leading personalities among the philosophers and the interpreters of the Hellenistic period. It was precisely due to his reputation that the court of the Attalids sent him to Rome, for a diplomatic mission at the time of Ennius' death, which occurred in 168 or 169 BC. Obviously, we are not interested in his activity as an author of grammatical, historical and philosophical writings, or in his debates, above all with the grammarian Aristarchus of Samothrace. What interests us instead is that by studying Homer and attaching to the *Iliad* hidden meanings with theories on the shape of the Earth and on the stretch of inhabited zones (which, obviously, could not be gathered from the shield of Achilles), he ended by attaching his own name to a "model" of Earth that has been named the "biblical-Cratesian synthesis"[75] which, as we shall see, will be held to be valid in all the Middle Ages.

---

[71]Horace, *Epistles*, II, 1, 156.

[72]See L. Russo: *The Forgotten Revolution: How Science was Born in 300 BC and Why It Had to Be Reborn* (Berlin: Springer, 2004).

[73]See, for instance, W. H. Stall: *Roman Science* (The University of Wisconsin Press, 1952), First Part.

[74]*Tusculanae Disputationes* I, 5: "Geometry also was in the highest esteem among them, and none were more illustrious than the mathematicians; while in this art we go no farther than is needful for the purpose of measuring and calculating" (Eng. trans. Andrew P. Peabody).

[75]See W. G. L. Randles: *De la terre plate au globe terrestre (Une mutation épistémologique rapide, 1480-1520)* (Paris: Librarie Armand Colin, 1980).

Crates imagined the Earth as a globe with four inhabited quarters, opposed to each other transversely and diagonally. They were separated by two oceanic strips, one of which (going from one pole to the other) divided the Eastern hemisphere from the Western one, and the other of which (around the equator) divided the Northern from the Southern hemisphere (see Fig. 1.5).

Crates adhered to Stoicism, which at that time also began to circulate in Rome, upon the arrival of Panaetius of Rhodes and his disciple Posidonius (135–51 BC), who afterwards also broadened to Platonic influences. We have already anticipated that Posidonius, in his turn, performed a measurement of the Earth. Posidonius cannot be defined a scientist: he was essentially a polymath and wrote many works (now all lost) on various subjects that met with great success in the Roman world. His measurement of the size of the Earth also gained notoriety. Let us explain the procedure of measurement. At the origin of all, there is the conviction (or the belief) that Rhodes and Alexandria (and therefore, by what we have already seen, Syene too) lie on the same meridian. Leaving from Rhodes when the star Canopus (modern name α Carinae) is on the horizon and sailing southward, one sees Canopus soaring into the sky. In Alexandria it appears at a maximum height of $7^\circ\ 30'$ over the horizon. This angle corresponds to the difference in latitude of the two towns (since they have been considered lying on the same meridian). Let us call this angle α (Fig. 1.6).

Posidonius starts from the consideration that the distance between Canopus and the Earth is much greater than the distance between Rhodes and Alexandria (what is, obviously, true) and thus that the lines from Canopus to Rhodes and that from Canopus to Alexandria can be considered parallel. Only the first line is drawn in Fig. 1.6 for reason of scale and for the same reason angle α is exaggerated and the distance Canopus-Earth reduced.

Considering the distance Rhodes-Alexandria equal to 5000 stades (it is curious that the distances contemplated in these measurements are always equal to 5000 stades), one has that the length of the terrestrial meridian is equal to $5000 \cdot \frac{360}{7.5} = 240{,}000$ stades.

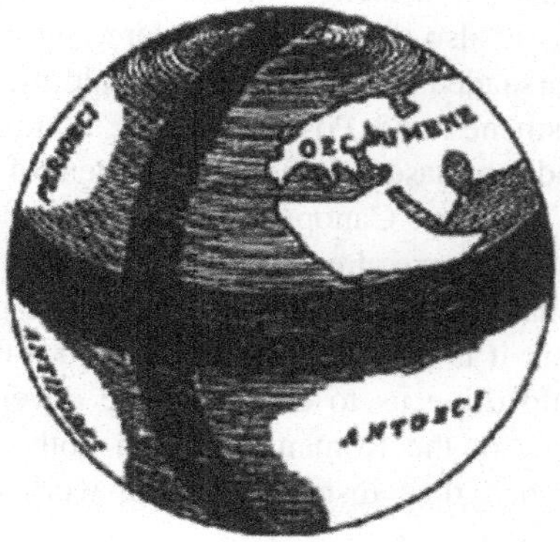

Fig. 1.5 A reconstruction of the globe of Crates. From Edward Luther Stevenson: *Terrestrial and celestial globes* (New Haven, 1921)

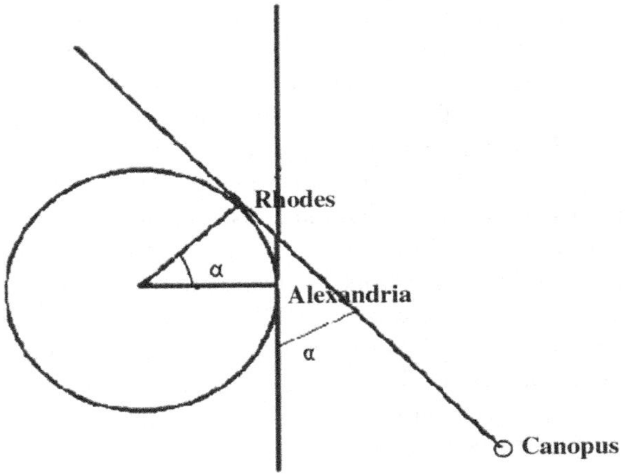

Fig. 1.6 Posidonius's measurement of the Earth

On the basis of what is known today, one can say that Posidonius' measurements were quite inaccurate:

(1) At the time of Posidonius, Canopus not only was visible on the horizon at Rhodes but reached a maximum elevation of more than 1° and remained on the horizon for a time of two hours and half;
(2) Rhodes and Alexandria have notably different longitudes;
(3) The measurement of 7.5° is also fairly inaccurate.

Of course, the problem of assigning a definite value to the stade still remains.

Several authors, both contemporary and posterior to Posidonius, report that Posidonius performed a second measurement and that this measurement obtained a remarkably inferior result. It is precisely the result of this measurement that was corroborated, centuries later, by Ptolemy and on the basis of which Columbus planned his voyage. We shall come back to this subject. For now, it is important for us to reiterate the fact that Eratosthenes' measurement was conceptually new and original (evidence of this is also that he is considered the founder of the scientific geodesy). We are not persuaded by the argument made by several authors that the two methods (of Eratosthenes and Posidonius) are comparable, justified on the grounds that both methods are based on the measurement of the declination of a star (in one case the Sun, in the other Canopus) and they are therefore equivalent. The novelty lies in the fact (as we have already said) that Eratosthenes does not perform astronomical observations, but only measurements on the ground.

At this point, we think it is convenient to depart from the chronological order, which in this case would drive us to deal with the development (or, better, the stagnation) of the science in the Roman world, in both the Republican and the Imperial period.[76] We shall deal instead with the work of Strabo (the greatest

[76] We shall treat this subject in the Chap. 2.

geographer of Antiquity) and of Ptolemy, whose works can be considered as a completion and conclusion of Greek astronomy and geography, after a vacuum in scientific research in the field that lasted about three centuries. These are works which have nothing to do with Roman culture; they are, in every way, late-blooming fruits of that splendid vine that was Greek culture.

1.8.1 *Strabo of Amasia (Pontus)*

Strabo (64 BC–24 AD), a native of Pontus, travelled widely, completed his education in Rome and lived in Alexandria. As we have already said, he is the most famous geographer of Antiquity; we could say he is the geographer par excellence. Seventeen books of his work *Geography* are extant. The first two of them (*Prolegomena*) deal with general subjects (the nature of geography, previous geographers, mathematical geography, shape and size of the Earth, mapping the Earth), and the remainder are dedicated to a description of the inhabited world of that time. Let us see what he claims in the incipit:

> The science of Geography, which I now propose to investigate, is, I think, quite as much as any other science, a concern of the philosopher; and the correctness of my view is clear for many reasons. In the first place, those who in earliest times ventured to treat the subject were, in their way, philosophers—Homer, Anaximander of Miletus, and Anaximander's fellow-citizen Hecataeus—just as Eratosthenes has already said; philosophers, too, were Democritus, Eudoxus, Dicaearchus, Ephorus, with several others of their times; and further, their successors—Eratosthenes, Polybius, and Poseidonius—were philosophers. In the second place, wide learning, which alone makes it possible to undertake a work on geography, is possessed solely by the man who has investigated things both human and divine—knowledge of which, they say, constitutes philosophy.[77]

The starting point (which we mentioned earlier in Sect. 1.1), is that he inserts Homer among the geographers; better, he consider him the first geographer.[78] As a matter of fact, Crates before him had maintained that Homer's poems were a source of geographical information. From Homer Strabo takes his model of the oecumene:

> In the first place, Homer declares that the inhabited world is washed on all sides by Oceanus and this is true; and then he mentions some of the countries by name, while he leaves us to infer the other countries from hints.[79]

Thus it is a sort of island encircled by the ocean. In fact, he goes on:

> We may learn both from the evidence of our senses and from experience that the inhabited world is an island; for wherever it has been possible for man to reach the limits of the earth,

[77]Strabo: *Geography* I, 1 (Eng. trans. H. Leonard Jones).

[78]See Francesco Prontera: "Sull'esegesi ellenistica della Geografia omerica" in *Geografia e Storia nella Grecia antica* (Florence: Leo S. Olschki Editore, 2011), pp 3–14.

[79]Strabo: *Geography* I, 1 (Eng. trans. H. Leonard Jones).

sea has been found, and this sea we call 'Oceanus'. And wherever we have not been able to learn by the evidence of our senses, there reason points the way.[80]

Before this he had said: "For he makes the sun to rise out of Oceanus and to set in Oceanus".[81]

As Aristotle had done earlier regarding the work of the philosophers who preceded him and as Ptolemy will do regarding the astronomers, so Strabo tells, summarizing and criticizing, the work of the geographers who dealt with tracing a picture of the oecumene before him. One he particularly criticises is Eratosthenes (also for the measurement of the Earth), whereas he has fewer objections to raise against Posidonius, who was also closer to him chronologically. This way of doing, however, has helped the scholars who, even if through a deforming lens, could collect information about the theories of the previous geographers. As an example, at the beginning of the second book of *Geography* we meet with an exposition of Eratosthenes' lost work, which begins in this way:

In the Third Book of his Geography Eratosthenes, in establishing the map of the inhabited world, divides it into two parts by a line drawn from west to east, parallel to the equatorial line; and as ends of this line he takes, on the west, the Pillars of Heracles, on the east, the capes and most remote peaks of the mountain-chain that forms the northern boundary of India. He draws the line from the Pillars through the Strait of Sicily and also through the southern capes both of the Peloponnesus and of Attica, and as far as Rhodes and the Gulf of Issus. Up to this point, then, he says, the said line runs through the sea and the adjacent continents (and indeed our whole Mediterranean Sea itself extends, lengthwise, along this line as far as Cilicia); then the line is produced in an approximately straight course along the whole Taurus Range as far as India, for the Taurus stretches in a straight course with the sea that begins at the Pillars, and divides all Asia lengthwise into two parts, thus making one part of it northern, the other southern; so that in like manner both the Taurus and the Sea from the Pillars up to the Taurus lie on the parallel of Athens.[82]

Strabo comes after about three centuries in which astronomy had been given a great boost in the Greek world, and the geocentric hypothesis, with an Earth of spherical shape, had been definitively adopted. The studies on spherical geometry (recall *On the moving Sphere* of Autolicus of Pitane) induced by astronomy have also provided with the scientific basis for representing the oecumene. As has been said, "the Greeks were led to know the Earth through the sky".[83]

Thanks to this, Eratosthenes introduced the system of coordinates. The data to be inserted (supplied by the experiences of the sailors and the accounts of the campaigns in the places where battles were fought) were few and not always exact. The network of meridians and parallels was conditioned to pass through known and important places and, as a consequence, the geometry was lopsided. Even so,

[80]Strabo: *Geography* I, 1 (Eng. trans. H. Leonard Jones).
[81]Strabo: *Geography* I, 1 (Eng. trans. H. Leonard Jones).
[82]Strabo: *Geography* II, 1 (Eng. trans. H. Leonard Jones).
[83]By the eminent Grecian scholar Giovanni Pugliese Carratelli.

Strabo had at his disposal a lot of material, elaborated after Eratosthenes also by Hipparchus, which he augmented with the information collected during his travels. Unlike the geography of Eratosthenes, that of Strabo was a descriptive geography without mathematical pretences.

As a matter of principle, he says, "For in many cases the way things appear to the sight and the agreement of all the testimony are more trustworthy than an instrument".[84] Moreover, regarding the mathematical geography of Eratosthenes, he says:

> For frequently Eratosthenes digresses into discussions too scientific for the subject he is dealing with, but, after he digresses, the declarations he makes are not rigorously accurate but only vague, since, so to speak, he is a mathematician among geographers, and yet a geographer among mathematicians; and consequently on both sides he offers his opponents occasions for contradiction.[85]

As we know, Eratosthenes was an important mathematician, esteemed by Archimedes, thus, Strabo's words are an unfair shaft of wit. But it has been used in the following centuries as a model to be applied according to the circumstances (we remember a case in the academic ambit in which the categories concerned were experimental physicists and theoretical physicists!).

1.8.2 Ptolemy's Geography

It is not known exactly either when Claudius Ptolemy, who lived in Alexandria in the second century AD, was born, or when he died (85–90 AD and 165–168 AD are considered as possible dates). The only known dates (those of his astronomical observations performed around the middle of the century, starting in 127 AD) are deducible from his principal work, the *Almagest*, which can be considered the swansong of the Greek astronomy. With Ptolemy, in fact, we have a momentary rebirth of the Greek astronomy after about two centuries of silence since the last of Hipparchus' observations.[86]

Ptolemy, who can be described as both a mathematician and astronomer and a geographer, is universally renowned for his mathematical model of the universe (precisely, the Ptolemaic system) but we only are interested in his work as a geographer, since our aim is to deal with the measurements and mathematical theories about the shape of the Earth. To this subject Ptolemy devoted a work: the *Introduction to Geography* (*Geographikè ufégesis*), usually referred to simply as *Geography*. Otto Neugebauer says of this work:

[84]Strabo: *Geography* II, 1 (Eng. trans. H. Leonard Jones).

[85]Strabo: *Geography* II, 1 (Eng. trans. H. Leonard Jones).

[86]On this, see L. Russo: *The forgotten revolution*, op. cit.

> Few books have exercised such a profound influence on human thought and civilization as Ptolemy's *Geography*. In its theoretical as well as practical consequences it far exceeds the importance of the astronomical "Ptolemaic System" and its Copernican modification, of interest and accessible to only a handful of men.[87]

However, this work had practically no followers in late Antiquity and the Middle Ages, and only became known in the Latin West starting in the fifteenth century. The only exceptions are Pappus of Alexandria, who about 300 AD wrote a description of the oecumene based on the work of Ptolemy, and the Roman historian Ammianus Marcellinus (fourth century) who used it; Cassiodorus (sixth century) mentioned it.

The *Geography*, as we have seen from the original title, was obviously written in Greek and perhaps for this reason was not read ("*graecum est, non legitur*" was the sentence used by the copyists of that time who could not copy the Greek manuscripts). At the end of the thirteenth century, the manuscript was found again by a Byzantine monk, Maximus Planudes, interested in rediscovering the Greek cultural heritage[88] and henceforth many copies were made. At the end of the fourteenth century a copy was brought to Italy by a Byzantine scholar, Emanuele Crisolora, invited to teach Greek in the University of Florence. A Latin translation undertaken by him was brought to its conclusion by his disciple Jacopo Angeli (or d'Angelo) da Scarperia (from 1405 to 1410), and was printed in 1475. Moreover, the Greek text was re-published by Erasmus in 1553 (Basel: Froben and Bischof).

Both the Latin and Greek texts contained errors (of conceptual nature and of typographical nature). In the sixteenth century an Italian translation of Girolamo Ruscelli was reprinted many times, and in 1828 a French translation by Abbé Nicholas Halma was published. An English translation published in 1992 by Edward Luther Stevenson is considered unreliable (O. Neugebauer considered the translator incompetent in technical matters[89]). At present there is no complete translation in a modern language considered reliable by scholars. For the Greek text, the edition by Karl Friedrich August Nobbe[90] and for the Latin text the (incomplete) edition of C. Müller and C. T. Fisher[91] are the references, whereas there is no complete translation like that of the *Almagest* due to G. J. Toomer.[92]

---

[87]O. Neugebauer: *A History of Ancient Mathematical Astronomy*, Part two (Springer-Verlag, 1975), p. 934.

[88]Maximus Planudes also reconstructed the 27 maps included in the work (see later) following (it seems) the instructions contained in Ptolemy's text. It is still under discussion if an original version of the 27 maps drawn by Ptolemy himself ever existed.

[89]See O. Neugebauer: "Ptolemy's Geography, Book VII, Chaps. 6 and 7". In: *Astronomy and History Selected Essays* (New York: Springer, 1983), p. 326.

[90]K. F. A. Nobbe: *Claudii Ptolemaei Geographia*, 3 vols. (Leipzig: Sumptibus et typis Caroli Tauchnitii, 1843–1945).

[91]C. Müller and C. T. Fisher (eds.): Claudius Ptolemaeus *Geographia*: Selections, 5 vols. (Paris: A. Firmin Didot, 1883–1901).

[92]*Ptolemy's Almagest*, trans. and annotations G. J. Toomer (Princeton University Press, 1998).

For the so-called theoretical chapters, there is the annotated translation of J. L. Berggren and A. Jones.[93]

Why have we gone into this long digression on the story of the editions of *Geography*? The answer is this: a present-day researcher who does not know ancient Greek but who wishes to study the Books 3–6, is obliged to untangle for himself the obscure passages of Ptolemy and the errors of the translators.

Let us now talk about the work, even if from our partial and limited point of view, focussing only the interdependence of the mathematical developments and the model proposed for the representation of the Earth. Ptolemy's *Geography* is the only Greek work on this subject which has been handed down to us from Classical Antiquity in a practically complete form. As he did with the *Almagest*, Ptolemy has, so to say, cannibalized the works of his predecessors. Only fragments and quotations of Eratosthenes, Hipparchus and Posidonius are extant. Ptolemy's immediate predecessor was Marinus of Tyre (not much older than Ptolemy himself), who flourished at the time of the emperor Trajan. All that we know about him is due to Ptolemy.

The *Geography* consists of eight books, the first of which explains the theoretical bases of the matter. Ptolemy even begins by defining geography (or, better, what he means by geography):

> [Geography] is the diagrammatic imitation of the known parts of the World with its unique features and it differs from Chorography since really this is the selecting out of certain regions as such to detail almost all the features in the smallest detail, and fixing in place such things as harbours, villages, towns and the course taken by rivers. The concern of [Geography] is to determine the nature of the Earth by showing it as one whole, how it is formed and, from one given point, show a comprehensive circumscription with contours, the location of rivers, great cities and races of people most worthy of mention, and the shape of every one of the most distinguished features.[94]

As to why it differs from chorography, he says,

> In its ultimate role, Chorography holds the key to describing just one part of the above mentioned whole as if one represented just the ear or the eye alone.
>
> But Geography is the viewing of the whole, the analogy being that concerned with showing the whole head.[95]

As a matter of fact, what Ptolemy calls "geography" would be more properly called "cartography in modern terms". The matter which Ptolemy is interested in, once the data at his disposal is assembled, is how to use them for drawing a map of the part of the world known at that time.

The conception Ptolemy has of the heavens, and consequently of the Earth, is the same he has already expressed in the *Almagest*. The cosmos consists of an immense sphere that rotates daily around an axis through its center where there is another

[93]John L. Berggren and Alexander Jones: *Ptolemy's Geography, An annotated translation of the theoretical chapters* (Princeton University Press, 2000).

[94]Ptolemy: *Geography* I, 1, 1–2 (Eng. trans. Louis Francis, who uses the term "Cartography" where we have substituted [Geography]).

[95]Ptolemy: *Geography* I, 1, 3 (Eng. trans. Louis Francis).

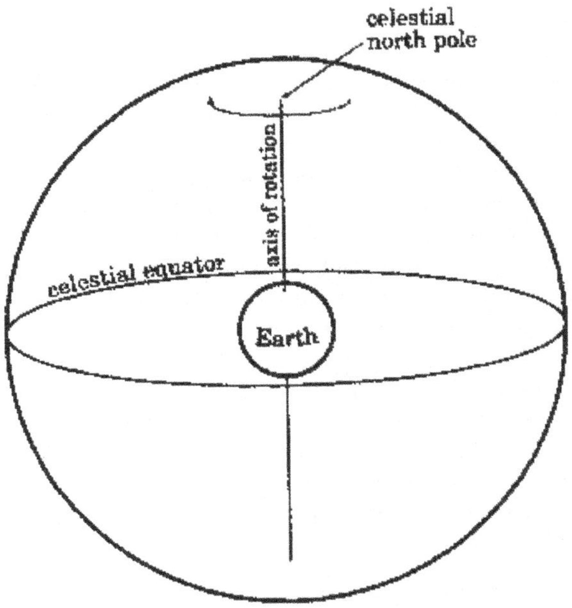

Fig. 1.7 The cosmos according to Ptolemy

(small) sphere, the Earth, which stands still.[96] The stars are fixed on the surface of the outer sphere. The two points of intersection of the axis of rotation with the sphere of the fixed stars are the north and south celestial poles. The daily rotation of the sphere of the fixed stars proceeds from east to west and clock-wise if seen from outside and above the north pole (see Fig. 1.7).

Further, the eight books of *Geography* were (or are thought to have been) supplemented by 27 maps, one general (of the known oecumene) and the other 26 of particular regions.

As we have mentioned above, Ptolemy talks about Marinus, often criticizing his results and methods, whereas Hipparchus is quoted (I, 4, 2) with esteem. He then enunciates the guidelines that one must follow for drawing a map of the oecumene:

> … before setting out to draw the inhabited World according to its true nature, we must comprehend with understanding and diligence previous scientific studies and the accounts of returning travellers. Such enquiries lead us to consider two categories of study; that of the geometry of the land itself and that of the meteorological study of the heavens. The latter is studied by an astrolabe, which is complex and dependent on the former, which is a gnomon, which is sufficient in itself.[97]

[96]Thomas S. Kuhn called this model “the two-sphere universe”; see *The Copernican revolution. Planetary Astronomy in the development of Western Thought* (Cambridge MA: Harvard University Press, 1957, Chap 1).

[97]*Geography* I, 2, 2 (Eng. trans. Louis Francis).

He first mentions the work of Marinus:

> Marinus the Tyrian, seems to be the last one we know of to pay attention to this matter in all sincerity and be honoured for it. For he brought to light complete reports that had been dispersed and remained unknown, and researched almost all the works of previous historians themselves, not only making good their errors but any earlier errors of his own in his geographical representations.[98]

He immediately goes on to list Marinus' errors: "First about legend and history, he claims that the Earth extends further eastward and southward than is demonstrably so".[99] According to Ptolemy, the oecumene extends in longitude more than in latitude. In longitude it extends 180° (12 h rather than 15 as asserted by Marinus) starting from the Fortunate Islands (the present-day Canaries) to China. In latitude one goes from 63° north (the parallel where Thule in supposed to be placed) to 16° 25′ south (the parallel of "anti-Meroë").

Once these boundaries are established, the data at hand must be adapted to a spherical surface and this transferred onto a map which, obviously, is a plane surface. From a theoretical point of view one must perform a projection from a sphere to a plane. If we know latitude and longitude of a locality on the sphere, we must give a rule of correspondence of it with a point on the map. We have used the adjective "theoretical" because latitude and longitude (resulting from astronomical measurements and for this considered reliable) were not always available.

Ptolemy elaborated two different projections (obviously, here, we are using a modern term unknown to Ptolemy). Let us explain the first one. As we have seen above, in Ptolemy's opinion the inhabited world (or, better, the world known at that time by the people living in the Roman Empire) covered only a part of the northern hemisphere, extending a little below the equator and having as a northern boundary the lands visited by Pytheas. Therefore, a simple conical projection able to reproduce meridians and parallels intersecting at right angles could be sufficient. The condition of orthogonality had also been satisfied by Marinus, who had drawn a rectangular grid which was, obviously, not altogether satisfactory for the northernmost part since the east-west distances came out exaggerated. Ptolemy retains the representations of meridians as straight lines converging to a point (the north pole) and the parallels as arcs of circles having the same centre (the north pole). The length of the parallel through Thule and of the equator are kept in the correct proportion to their true magnitudes. For the rest, one parallel (that through Rhodes was considered the most appropriate) was assumed as a reference and the ratio 5:4 was chosen for the spacing of the lines separated by a given number of degree of latitude to that of the lines separated by the same numbers of degrees of longitude.

In this way, the spacing of places north of the latitude of Rhodes turns out to be contracted and that of places south of Rhodes expanded. For Ptolemy the distortion was too greatly exaggerated for the parallels south of the equator, since in the reality they go shortening southward. He rectified the distortion by drawing the arcs of

[98]*Geography* I, 6, 1 (Eng. trans. Louis Francis).

[99]*Geography* I, 6, 3 (Eng. trans. Louis Francis).

parallels south of the equator of the same length of their symmetrical correspondents north of the equator. The segments of meridians south of the equator are obtained by joining the points of the same longitude of the corresponding parallels (see Fig. 1.8).

Thus south of the equator the projection is pseudoconical. This last modification obviously forbids one to locate the points south of the equator with the same method that can be used in the part north of the equator, i.e., by making use of a ruler hinged in the point representing the north pole, as it was a rotating meridian.

The same difficulty arises for the second map (see Fig. 1.9). In it, Ptolemy imagines the oecumene as being observed by one who looks at its central part from afar. Thus, the central meridian appears as rectilinear, whereas the others appear as arcs of a circle with their concavities facing the central meridian. The parallels again are concentric circles: three of them (the parallel of Thule, that of Syene and that of anti-Meroë) are chosen to be of lengths proportional to their true lengths. On the parallels one marks the points corresponding to intervals of five degrees and then joins triples of corresponding points with circular arcs (a circular arc can always be drawn through any three noncollinear points): these represent the meridians.

In this way one obtains a grid which can be complemented by drawing all wanted parallels. In this map the correspondence between the points on the sphere and their image on the plane surface is much more faithful than that in the first map. But the drawing of the map turns out to be more arduous, as it no longer has rectilinear meridians.

Both maps can be classified (in modern terms) as pseudoconical. In any case, Ptolemy did not exploit them for drawing the 26 regional maps, for which he limited himself to using the cylindrical projection, that is, a rectangular grid like that of Marinus.

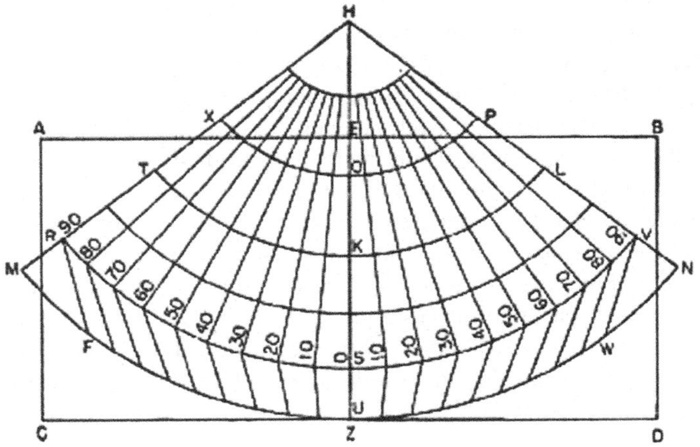

Fig. 1.8 Ptolemy's segments of meridian south of the equator

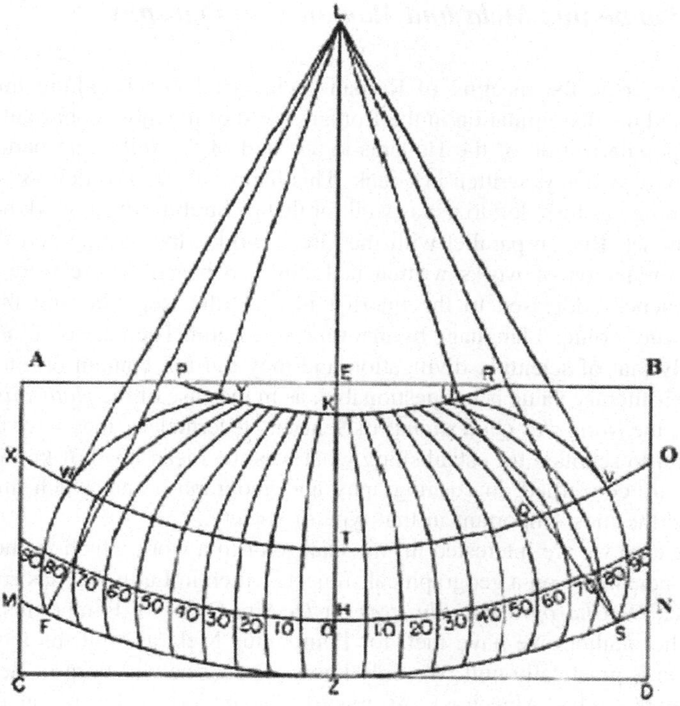

Fig. 1.9 The oecumene as observed by one south of the equator

Coming back, now, to talk about the work as a whole, there are three remarks which must be made. Firstly, the work contains many errors in the fixing of coordinates of localities, owing both to incorrect information and inaccurate observations. Secondly, there is a background error due to the fact that Ptolemy accepted, and in so doing confirmed, the measure of the terrestrial meridian used by his predecessor Marinus. Finally, the longitudes were affected by a systematic error which dilated their differences. There is a copious literature on this; we limit ourselves to referring the reader to the recent paper by Lucio Russo.[100]

In any case, one cannot not agree with Otto Neugebauer when he says:

> It is in the *Geography* that for the first time a mathematically clear theory of geographical mapping was presented and with it went the creation of a consistent grid of coordinates, reckoned in degrees.[101]

---

[100] L. Russo: *Ptolemy's longitudes*, loc. cit.

[101] O. Neugebauer: *A History of the ancient mathematical Astronomy*, op. cit., p. 934.

1.8.3 Pomponius Mela and Roman Cosmography

As we have seen, the meeting of Roman society and Greek culture implied the presence and the dissemination in the Roman world of the most important works of scientific popularization of the Hellenistic age and of the following period. These works were, obviously, written in Greek. This happened for Posidonius' works (of which nothing has been left to us) as well for that of Strabo, whose work has luckily survived intact. But, in parallel with the Greek works, the writing and the consequent dissemination of works written in Latin also began. These were works of popular science addressed to the internal market (the Republic and the Roman Empire), whose official language by that time was Latin. The nature of those works was merely that of scientific divulgation and they did not contain original results, even if their literary value is unquestionable, as in the case of the *Naturalis Historia* of Pliny or the *Naturales Quaestiones* of Seneca. Nevertheless, they have turned out to be useful to scholars for establishing what was the mean level of knowledge (as far as we are concerned, in cosmography and geography) and which things were considered the most important in that type of society.

For the case we are interested in, we shall refer to a work which is undoubtedly the most ancient one on a geographical subject written in Latin that has come down to us intact: *De Chorographia libri tres* (or *De Situ Orbis*) of Pomponius Mela. As for the other authors we have met, for Pomponius Mela as well the biographical information is practically null. We only know for sure he was born in Betic Spain, in Tingentera (today Algeciras). As regards the period, a discussion has arisen among scholars from the interpretation of a passage of the work (III, 49) which tells about a battle fought in Britannia and won by a "*Principium maximus*". The interpretation almost universally accepted is that the "*princeps*" involved is the emperor Claudius, in which case the work should have been written between the end of 43 and the beginning of 44 AD.[102] Further, from the interpretation of a passage of the *Proemium* ("*Dicam autem alias plura et exactius, nunc ut quaeque erunt clarissima et strictim*"[103]) arose the obvious conclusion that the author announced the writing of a more extensive work, of which this was an epitome. Anyway, the fact is that either the work was never written or it has been lost; moreover, there is no trace of its existence in the writings of subsequent authors.

Let us pass on to the structure of the work. After a short introduction of general character (about which we shall comment on later), in the first book Mela, starting

[102]For this and the other discussion regarding *De Chorographia*, we refer the reader to *Pomponii Melae De Chorographia libri tres*. Introduction, critical edition and commentary by Piergiorgio Parroni (in Italian) (Rome: Edizioni di Storia e Letteratura, 1984).

[103]*Pomponii Melae De Chorographia libri tres*, ed. C. Frick (Leipzig: Teubner, 1880), p.1 (rpt. Stuttgart: W. Schaub, 1967), p. 1. Eng. trans: "I should, however, say more elsewhere and with greater preciseness" (*Pomponius Mela's Description of the World*, trans. Frank E, Romer (University of Michigan Press, 1998, p. 33).

from the Straits of Gibraltar (*Gaditanum fretum*), describes the regions of Africa and Asia which look out on the Mediterranean. In the second, he describes Europe, coming back from west to east to the starting point, and the islands of the Mediterranean. Finally, in the third book, he describes the regions on the Ocean, from the Straits of Gibraltar to the Baltic sea and from the Caspian coming again, through the south, to Gibraltar.

Only chapter IV of Book II deals with Italy. Mela justifies this fact by saying that Italy is known to everyone. In contrast, we recall that Strabo addressed his Books V and VI to Italy. In addition, we also remark that Mela never quotes Strabo, who had finished his work about twenty years before. In this he is followed by Pliny and even by Ptolemy. Seneca, instead, quotes Posidonius in his *Naturales Quaestiones*.

Let us now come to the initial part of the first chapter of the first book, where Mela enunciates (but the verb is excessive) the principles of cosmography he follows. Before all, we remark that Mela laments that the subject he will deal with is not suitable for writing a work of eloquence: "*Orbis situm dicere aggredior, impeditum opus, et facundiae minime capax*".[104] In any case, he gives a very succinct description of the world:

> [3] Whatever all this is, therefore, on which we have bestowed the name of world and sky, it is a single unity and embraces itself and all things with a single ambit. It differs in its parts. Where the sun rises is designated formally as east or sunrise; where it sinks, as west or sunset; where it begins its descent, south; in the opposite direction, north. [4] In the middle of this unity the uplifted earth is encircled on all sides by the sea. In the same way, the earth also is divided from east to west into two halves, which they term hemispheres, and it is differentiated by five horizontal zones. Heat makes the middle zone unlivable, and cold does so to the outermost ones. The remaining two habitable zones have the same annual seasons, but not at the same time. The Antichtones inhabit one, we the other. The chorography of the former zone is unknown because of the heat of the intervening expanse, and the chorography of the latter is now to be described. [105]

By the way, his readership, i.e., the Romans of that time, was most probably not very interested in either cosmography or the theories regarding the shape of the Earth: a brief mention of Crates' model with the five zones was more than enough for people mainly interested in traveling overland.

[104] *De Chorographia*, op. cit., p. 1: "A description of the known world is what I set out to give, a difficult task and one hardly suited to eloquence..." (ibidem).

[105] *De Chorographia*, op. cit., pp 1–2: "*Omne igitur hoc, quidquid est cui mundi caelique nomen indidimus, unum id est et uno ambitu se cunctaque amplectitur. partibus differt; unde sol oritur oriens nuncupatur aut ortus, quo demergitur occidens vel occasus, qua decurrit meridies, ab adversa parte septentrio. huius medio terra sublimis cingitur undique mari, eodemque in duo latera quae hemisphaeria nominant ab oriente divisa ad occasum zonis quinque distinguitur. mediam aestus infestat, frigus ultimas; reliquae habitabiles paria agunt anni tempora verum non pariter. antichthones alteram, nos alteran incolimus. illius situs ob ardorem intercedentis plagae incognitus, huius dicendus est*" (Eng. trans. *Pomponius Mela's Description of the World*, trans. Frank E. Romer (University of Michigan Press, 1998, p. 34).

Sad to say, in spite of the blaze represented by Ptolemy's *Geography*, almost two centuries later, it will be Crates' model, burdened by the provision of the Bible, that will intrigue the encyclopedists who worked in the last centuries of the Roman Empire and the early Middle Ages.

Suggested Readings

Courie, D. L. (2011). *Heaven and earth in ancient Greek cosmology: From Thales to Heraclides Ponticus*. Berlin: Springer.

Martin, T. R. (2013). *Ancient Greece: From prehistoric to Hellenistic Times* (2nd ed.). New Haven: Yale University Press.

Farrington, B. (2000). *Greek science*. With an introduction of Joseph Needham: Spokesman Books.

Heath, T. (1981). *A history of Greek mathematics*. Oxford: Clarendon Press (1921) (Rpt. New York: Dover 1981).

Kahn, C. H. (2001). *Pythagoras and the Pythagoreans: A brief history*. Indianapolis: Hacket Publishing Company.

Chapter 2
The Roman World from the End of the Republic to the End of the Empire

In dealing with our subject, as far as the Roman world is concerned, we are obliged to depart from the rigorous chronological order. Many times one has the impression that parallel worlds coexisted and did not communicate, at least from the viewpoint of the scientific information. This is certainly not due to the difficulties of communication: despite of the rudimentary technologies concerning communication media, in the third century BC Eratosthenes (in Alexandria) and Archimedes (in Syracuse) shared information on their progress in mathematical research. One can also think that the inner circle of learned persons in the Roman society was not equally inclined to accept the new theories and their relevant evidences. We have an example of this in the case of Lucretius.

We have already seen that Ptolemy's *Geography* was ignored in the Latin West for at least eleven centuries. It was a bottled message re-emerged from the sea of the history firstly in Byzantium and then transferred in the Latin West. We shall see now how bumpy the road of the model of a spherical Earth was in the nearly five centuries between the end of the Republic and the end of the Empire. A disruptive event which influenced the ideas about the shape and the constitution of the Earth, as we shall see, was the spread and the subsequent institutionalization of the Christian doctrine.

2.1 The Age of Augustus

Posidonius (135–51 BC), about whom we have already spoken, was undoubtedly the most influential intellectual of his age in the Roman world, even though he wrote in Greek. His numerous works, all lost, were copied by the Roman writers in subsequent years, starting with Pliny and Seneca and into the Middle Ages.

Cicero (106–43 BC) belonged to the generation succeeding that of Posidonius and, besides being acquainted with him, drew inspiration from his work when he needed to express a cosmographical model. One of the most important works of

D. Boccaletti, *The Shape and Size of the Earth*,
https://doi.org/10.1007/978-3-319-90593-8_2

Cicero is *Republic*, which takes inspiration from the homonymous work of Plato. Almost the entire text of the work was lost,[1] perhaps a little after 1000 AD, and only its final part survived, the so-called *Somnium Scipionis*.

The *Republic* appears as a dialogue of political doctrine. In it, Cicero projects himself into the past in order to find the best form of state in the Roman constitution at the time of the Scipions. The principal interlocutors in the dialogue are Scipio Aemilianus and Laelius. Book VI ended with the so-called *Somnium Scipionis*, in which Scipio Aemilianus recalls the dream in which his grandfather Scipio Africanus showed him from on high the littleness of worldly things and the beatitude that awaited great statesmen in the hereafter. Cicero being an eclectic, elements of different philosophies (Platonic, Pythagorean, Stoic) merge in his representation of the hereafter.

The *Somnium* has been studied, both for its philosophical content, and as a masterpiece of style, since it was considered one of the most beautiful passages of Latin prose. For these reasons it survived even when the rest of the work disappeared. In fact, commentaries on this writing began to appear very soon. We shall speak below about one of them (that of Macrobius). For now, we are only interested in the passages where the cosmographical model held by Cicero is enounced. Let us look at three of them. In the dream, while Scipio Aemilianus continues to observe the Earth still with greater attention, the grandfather intervenes:

> … how long, I pray you, (said Africanus) will your mind be fixed on that object—why don't you rather take a view of the magnificent temples whither you have arrived? The universe is composed of nine circles or rather spheres, one of which is the most elevated, and is exterior to all the rest which it embraces; and where the Supreme God resides, who bounds and contains the whole. In it are fixed those stars which revolve with never–varying courses. Below this are seven other spheres, which revolve in a contrary direction to that in the heavens. One of these is occupied by the globe, which on earth they call Saturn. Next to that, is the star of Jupiter, so benign and salutary to mankind. The third in order, is that fiery and terrible planet called Mars. Below this again, almost in the middle region, is the Sun,—the leader, governor, and prince of the other luminaries; the soul of the world, which it regulates and illumines, filling all things with its rays. Then follow Venus and Mercury, which attend as it were on the Sun. Lastly, the Moon, which shines only in the reflected beams of the Sun, moves in the lowest sphere of all. Below this, if we except that gift of the gods, human souls, every thing is mortal, and tends to dissolution, but above it all is eternal. For the Earth, which is the ninth globe, and occupies the centre, is immoveable, and being the lowest, all others gravitate towards it.[2]

[1]A considerable part of it was recovered by the Italian philologist Angelo Mai in December 1819.

[2]*De Republica*, VI, 17: *Quaeso," inquit Africanus, "quousque humi defixa tua mens erit? Nonne aspicis, quae in templa veneris? Novem tibi orbibus vel potius globis conexa sunt omnia, quorum unus est caelestis, extimus, qui reliquos omnes complectitur, summus ipse deus arcens et continens ceteros; in quo sunt infixi illi, qui volvuntur, stellarum cursus sempiterni. Cui subiecti sunt septem, qui versantur retro contrario motu atque caelum. Ex quibus summum globum possidet illa, quam in terris Saturniam nominant. Deinde est hominum generi prosperus et salutaris ille fulgor, qui dicitur Iovis; tum rutilus horribilisque terris, quem Martium dicitis; deinde subter mediam fere regionem Sol obtinet, dux et princeps et moderator luminum reliquorum, mens mundi et temperatio, tanta magnitudine, ut cuncta sua luce lustret et compleat. Hunc ut comites consequuntur Veneris alter, alter Mercurii cursus, in infimoque orbe Luna radiis solis accensa convertitur. Infra*

At this point, we can remark that Cicero also indulges in astrological sentences. In addition, we note that in ordering the celestial bodies, he follows Posidonius assigning to the Sun a central position, while Plato, following the Pythagoreans, put it in the sixth position between Mercury and the Moon.

In the dream, Aemilianus also hears music and asks the grandfather, who explains:

> It is that which is called the music of the spheres, being produced by their motion and impulse; and being formed by unequal intervals, but such as are divided according to the justest proportion, it produces, by duly tempering acute with grave sounds, various concerts of harmony. For it is impossible that motions so great should be performed without any noise; and it is agreeable to nature that the extremes on one side should produce sharp, and on the other, flat sounds. For which reason the sphere of the fixed stars, being the highest, and carried with a more rapid velocity, moves with a shrill and acute sound; whereas that of the moon, being the lowest, moves with a very flat one. As to the Earth, which makes the ninth sphere, it remains immoveably fixed in the middle or lowest part of the universe. But those eight revolutionary circles, in which both Mercury and Venus are moved with the same celerity, give out sounds that are divided by seven distinct intervals, which is generally the regulating number of all things.[3]

After this, since Aemilianus is still employed in contemplating the seat and residence of mankind, Africanus continues:

> The earth, you see, is peopled but in a very few places, and those too of small extent; and they appear like so many little spots of green, scattered through vast uncultivated deserts. Its inhabitants are not only so remote from each other as to cut off all mutual correspondence; but their situation being in oblique or contrary parts of the globe, or perhaps in those diametrically opposite to yours, all expectations of universal fame must fall to the ground. ...You may likewise observe that the same globe of the earth is girt and surrounded with certain zones, whereof those two that are most remote from each other, and lie under the opposite poles of heaven, are congealed with frost; but that one in the middle, which is far the largest, is scorched with the intense heat of the sun. The other two are habitable, one towards the south,—the inhabitants of which are your Antipodes, with whom you have no connection;—the other, towards the north, is that you inhabit, whereof a very small part, as you may see, falls to your share. For the whole extent of what you see, is as it were but a

---

autem iam nihil est nisi mortale et caducum praeter animos munere deorum hominum generi datos; supra Lunam sunt aeterna omnia. Nam ea, quae est media et nona, Tellus, neque movetur et infima est, et in eam feruntur omnia nutu suo pondera (Eng. trans. Francis Barham).

[3]*De Republica*, VI, 18: "*Hic est, qui intervallis disiunctus imparibus, sed tamen pro rata parte distinctis, impulsu et motu ipsorum orbium efficitur et acuta cum gravibus temperans varios aequabiliter concentus efficit; nec enim silentio tanti motus incitari possunt, et natura fert, ut extrema ex altera parte graviter, ex altera autem acute sonent. Quam ob causam summus ille caeli stellifer cursus, cuius conversio est concitatior, acuto et excitato movetur sono, gravissimo autem hic lunaris atque infimus; nam terra nona immobilis manens una sede semper haeret complexa medium mundi locum. Illi autem octo cursus, in quibus eadem vis est duorum, septem efficiunt distinctos intervallis sonos, qui numerus rerum omnium fere nodus est* (Eng. trans. Francis Barham).

little island, narrow at both ends and wide towards the middle, which is surrounded by the sea, which on earth you call the great Atlantic Ocean, and which, notwithstanding this magnificent name, you see is very insignificant.[4]

Contemporaneous and friend of Cicero was Marcus Terentius Varro (116–27 BC), a great and influential intellectual, known for the great number of works (74 in 620 books), practically all lost except for the small work *De Agricultura* (3 books), written when he was 80 years old, and a part (6 books) of the great work *De lingua Latina*. We mention Varro, even if among his extant writings there are no excerpts regarding our subject, in order to remark once again (as happened before for Posidonius) how the widespread tendency of Roman society of the time to prefer the handbooks and the popular works of small size caused many important works, once they had been summarized by less important writers, to disappear.

Now, we go on to deal with a great poet, Titus Lucretius Carus, also a friend of Cicero. The date of birth of Lucretius (like that of many others) is not certain (it falls in any case in the first decade of the first century BC), while for the death we know the date of 55 BC. The scanty biographical notes are provided in the *Chronicon* (380 AD) by St. Jerome (347–419 AD), who translated the Bible (387 AD) into Latin from Greek and Hebrew.[5] According to scholars, it seems that Jerome took these notes from a lost work of Suetonius and that they are not completely believable (for instance, that he was driven mad by a love-potion). Jerome also asserts that Cicero "*emendavit*" Lucretius' work, i.e., in modern terms, that he was the editor of *De rerum natura*, a poem in 6 books.

As the title (quite usual in Antiquity) indicates, this grand poem deals with the universe in all its components. In Antiquity this work, after a short period, also fell into oblivion and was rediscovered by the Italian humanist Poggio Bracciolini in 1417. In it, Lucretius exposes the philosophical creed of Epicurus, with the atomistic theory of Leucippus and Democritus. For the motion of atoms, Lucretius introduces the theory of *clinamen*, i.e., the declination from the vertical of the atoms

[4]*De Republica*, VI, 20, 21: *Vides habitari in terra raris et angustis in locis et in ipsis quasi maculis, ubi habitatur, vastas solitudines interiectas eosque, qui incolunt terram, non modo interruptos ita esse, ut nihil inter ipsos ab aliis ad alios manare possit, sed partim obliquos, partim transversos, partim etiam adversos stare vobis; a quibus exspectare gloriam certe nullam potestis.*

Cernis autem eandem terram quasi quibusdam redimitam et circumdatam cingulis, e quibus duos maxime inter se diversos et caeli verticibus ipsis ex utraque parte subnixos obriguisse pruina vides, medium autem illum et maximum solis ardore torreri. Duo sunt habitabiles, quorum australis ille, in quo, qui insistunt, adversa vobis urgent vestigia, nihil ad vestrum genus; hic autem alter subiectus aquiloni, quem incolitis, cerne quam tenui vos parte contingat! Omnis enim terra, quae colitur a vobis, angustata verticibus, lateribus latior, parva quaedam insula est circumfusa illo mari, quod "Atlanticum", quod "magnum", quem "Oceanum" appellatis in terris; qui tamen tanto nomine quam sit parvus, vides (Eng. trans. Francis Barham).

[5]Jerome: *Chronicon, : Titus Lucretius poeta nascitur, postea amatorio poculo in furorem versus cum aliquot libros per intervalla insaniae conscripsisset quos postea Cicero emendavit, propria se manu interfecit anno aetatis XLIIII* (The poet Titus Lucretius is born. Later, having become insane by drinking a love potion, after writing during period of remission several books, which Cicero later edited, he committed suicide at the age of forty-four) (Eng. trans. Joseph Farrell).

which fall under their weight. It is the *clinamen* which allows the atoms to associate and constitute the various kinds of matter.

Some modern authors (among them, for instance, Michel Serres and Piergiorgio Odifreddi[6]) have seen *in nuce* in *De rerum natura* some of modern physical theories. For our part, we are obliged to remark that the allegiance to the theories of Epicurus (342–270 BC), at an interval of time of more than two centuries, forces Lucretius to reject the theory of the spherical shape of the Earth, already accepted at that time by learned Roman society.

We report below an excerpt from the first book, which begins with an invocation to Venus, symbol of the vital principle, where the sphericity of the Earth is violently denied:

> And in these problems, shrink, my Memmius, far
> from yielding faith to that notorious talk:
> that all things inward to the centre press;
> and thus the nature of the world stands firm
> with never blows from outward, nor can be
> nowhere disparted since all height and depth
> have always inward to the centre pressed
> (if thou art ready to believe that aught
> itself can rest upon itself); or that
> the ponderous bodies which be under earth
> do all press upwards and do come to rest
> upon the earth, in some ways upside down,
> like to those images of things we see
> at present through the waters. They contend,
> with like procedure, that all breathing things
> head downward roam about, and yet cannot
> tumble from earth to realms of sky below,
> no more than these our bodies wing away
> spontaneously to vaults of sky above;
> that, when those creatures look upon the sun,
> we view the constellations of the night;
> and that with us the seasons of the sky

[6]Michel Serres: *La naissance de la physique dans le texte de Lucrèce* (Les Editions de Minuit, 1998); Piergiorgio Odifreddi: *Come stanno le cose. Il mio Lucrezio, la mia Venere* (Rizzoli, 2013).

they thus alternately divide, and thus
do pass the night coequal to our days.[7]

This passage which comes after a part dedicated to the formation of matter, is certainly considered important by the poet since he addresses his friend Memmius to whom the poem is dedicated. Further on, he continues:

Nor is there any place, where, when they've come,
bodies can be at standstill in the void,
deprived of force of weight; nor yet may void
furnish support to any,- nay, it must,
true to its bent of nature, still give way.
Thus in such manner not at all can things
be held in union, as if overcome
by craving for a centre.[8]

Finally, in Book V (dedicated to Epicurus), a curious explanation of the stability of the flat disc-shaped Earth reads:

And that the earth may there abide at rest
in the mid-region of the world, it needs
must vanish bit by bit in weight and lessen,
and have another substance underneath,
conjoined to it from its earliest age
in linked unison with the vasty world's
realms of the air in which it roots and lives.
On this account, the earth is not a load,
nor presses down on winds of air beneath;
even as unto a man his members be
without all weight – the head is not a load

[7]Lucretius: *De rerum natura*, I, 1052–1067: *Illud in his rebus longe fuge credere, Memmi, /in medium summae quod dicunt omnia niti /atque ideo mundi naturam stare sine ullis /ictibus externis neque quoquam posse resolvi /summa atque ima, quod in medium sint omnia nixa, /ipsum si quicquam posse in se sistere credis, /et quae pondera sunt sub terris omnia sursum /nitier in terraque retro requiescere posta, /ut per aquas quae nunc rerum simulacra videmus; /et simili ratione animalia suppa vagari /contendunt neque posse e terris in loca caeli /reccidere inferiora magis quam corpora nostra /sponte sua possint in caeli templa volare; /illi cum videant solem, nos sidera noctis /cernere et alternis nobiscum tempera caeli/ dividere et noctes parilis agitare diebus* (Eng. trans. William Ellery Leonard).

[8]Lucretius: *De rerum natura*, I, 1077–1082: *Nec quisquam locus est, quo corpora cum venerunt, / ponderis amissa vi possint stare <in> inani; /nec quod inane autem est ulli subsistere debet, /quin, sua quod natura petit, concedere pergat. /haud igitur possunt tali ratione teneri /res in concilium medii cuppedine victae* (Eng. trans. William Ellery Leonard).

unto the neck; nor do we feel the whole
weight of the body to centre in the feet.
But what so weights come on us from without,
weights laid upon us, these harass and chafe,
though often far lighter. For to such degree
it matters always what the innate powers
of any given thing may be. The earth
was, then, no alien substance fetched amain,
and from no alien firmament cast down
on alien air; but was conceived, like air,
in the first origin of this the world,
is a fixed portion of the same, as now
our members are seen to be a part of us.[9]

2.2 From Augustus to the Age of Diocletian

In the age of Augustus (63 BC–14 AD) several poets flourished and their appreciation of the poetic work of Lucretius was unanimous. For instance, Ovid, in his work *Amores*, says: "Then, the works of sublime Lucretius will endure".[10]

The appreciation continued until the third century also from the Christian writers who, though fighting against the Epicurean doctrine which some of them (as, for instance, Lactantius) had even grasped from reading of *De Rerum Natura*, praised its poetry. Vitruvius (about whom we shall talk below) also recalls the immediateness of his poetic message: "So, too, numbers born after our time will feel as if they were discussing nature face to face with Lucretius".[11]

---

[9]Lucretius: *De rerum natura*, V, 534–549: *Terraque ut in media mundi regione quiescat, /evanescere paulatìm et decrescere pondus /convenit atque aliam naturam supter habere /ex ineunte aevo coniunctam atque uniter aptam /partibus aëriis mundi, quibus insita vivit. /Propterea non est oneri neque deprimit auras, /ut sua cuique nomini nullo sunt pendere membra, /nec caput est oneri collo, nec denique totum; /corporis in pedibus pondus sentimus inesse; /at quae cumque foris veniunt inpostaque nobis /pondera sunt laedunt, permulto saepe minora. /Usque adeo magni refert quid quaeque queat res. /sic igitur tellus non est aliena repente /allata atque auris aliunde obiecta alienis, /sed pariter prima concepta ab origine mundi /certaque pars eius, quasi nobis membra videntur* (Eng. trans. William Ellery Leonard).

[10]Ovid: *Amores* I, 15, 23: "*carmina sublimis tunc sunt peritura Lucreti*" (Eng. trans. A. S. Kline).

[11]Vitruvius: *De Architectura* IX, 17: *Plures post nostram memoriam nascentes cum Lucretio videbuntur velut coram de rerum natura disputare* (Eng. trans. M. H. Morgan).

Apart from Strabo, who fully belongs to the Augustan age (indeed, he is practically coetaneous with August) but writes in Greek and is representative of the Greek culture, we do not meet in this age outstanding authors dealing with astronomy or geography to whom we can refer to obtain authoritative testimony of the conventional wisdom regarding the Earth. With scant diplomacy, Strabo wrote:

> Now although the Roman historians are imitators of the Greeks, while the fondness for knowledge that they of themselves bring to their histories is inconsiderable; hence, whenever the Greeks leave gaps, all the filling in that is done by the other set of writers is inconsiderable —especially since most of the very famous names are Greek.[12]

Our aim is not only to understand, and support with documentary evidence, what the points of view of the scientists, or in any case of the outstanding thinkers, in the various ages was about the shape of the Earth and related problems. It is also important for us to know what, generally speaking, the ruling élite, in a certain country, thought in a certain time. At present we are interested in the first century AD and, in this age, the most important culture, and the one whose public opinion is the reference par excellence is the Roman Empire.

The fact that the Romans did not exhibit a marked interest in theoretical science, in either the case of astronomy or astronomical geography, is a guarantee that what one can read in the works of historians and polymaths is really a reflection of the conventional wisdom: bad and good is what learned society thought.

About a generation after Lucretius we meet another great poet, Virgil (70–19 BC), the most important poet of the age of Augustus. He never dealt with the subject of our interest; he was interested in the earth but not in the Earth:

> Me before all things may the Muses sweet,
> Whose rites I bear with mighty passion pierced,
> Receive, and show the paths and stars of heaven,
> The sun's eclipses and the labouring moons,
> From whence the earthquake, by what power the seas
> Swell from their depths, and, every barrier burst,
> Sink back upon themselves, why winter-suns
> So baste to dip 'neath ocean, or what check
> The lingering night retards.
> But if to these high realms of nature the cold curdling blood
> About my heart bar access, then be fields
> And stream-washed vales my solace, let me love
> Rivers and woods, inglorious.[13]

---

[12]Strabo: *Geography* III, 4, 116 (Eng. trans. Horace Leonard Jones).

[13]Virgil: *Georgicon*, II, 475–486: *Me vero primum dulces ante omnia Musae, /quarum sacra fero ingenti percussus amore, /accipiant caelique vias et sidera monstrent, /defectus solis varios lunaeque labores; /unde tremor terris, qua vi maria alta tumescant /obicibus ruptis rursusque in*

However, in a—so to speak—indirect way we can find what we are looking for in the work *De Architectura* by Vitruvius (Marcus Vitruvius Pollio, about 80–15 BC). The dates we have reported correspond to the opinion generally accepted nowadays (another opinion placed Vitruvius' work as late as the fourth century AD), which also fixes the publication of the work, *inter alia* dedicated to Augustus, about the penultimate decade of the first century BC. Thus, we can say that *De Architectura* is the first of the works of the Imperial period to which we can refer for the subject we are interested in.

The work consists of ten books and deals, albeit succinctly, with all subjects which concern an architect or which have something to do with buildings, machines, instruments, etc. According to the opinion of scholars, this work is a compilation which exploits as sources Hellenistic works (whether directly or through Roman works traceable to them). The ninth book is the one dedicated to astronomy and related subjects. As Vitruvius himself says in the introduction to that book:

> I have written the present books, in the first seven treating of buildings and in the eighth of water. In this I shall set forth the rules for dialling, showing how they are found through the shadows cast by the gnomon from the sun's rays in the firmament, and on what principles these shadows lengthen and shorten.[14]

The book begins with a peroration. Vitruvius asks why, if the wrestlers in ancient Greece who were victorious in Olympia were glorified, are philosophers such as Pythagoras, Democritus, Plato, Aristotle not glorified in the same way? He writes:

> Since, therefore, individuals as well as the public are so indebted to these writers for the benefits they enjoy, I think them not only entitled to the honour of palms and crowns, but even to be numbered among the gods. I shall produce in illustration, some of their discoveries as examples, out of many, which are of utility to mankind, on the exhibition whereof it must be granted without hesitation that we are bound to render them our homage.[15]

Vitruvius' exposition of planetary motion contains some inexactitudes and obscurities. Moreover, it seems he accepts both the theory of Heraclides (Mercury and Venus rotating around the Sun) as well as that of the planetary ordering (the

---

se ipsa residant, /quid tantum Oceano properent se tingere soles /hiberni, vel quae tardis mora noctibus obstet. /sin has ne possim naturae accedere partis /frigidus obstiterit circum praecordia sanguis, /rura mini et rigui placeant in vallibus amnes, /flumina amem silvasque inglorious (Eng. trans. J. B. Greenough).

[14] Vitruvius: *De Architectura* IX, Praefactio, 18: *ea volumina conscripsi, et prioribus septem de aedificiis, octavo de aquis, in hoc de gnomonicis rationibus quemadmodum de radiis solis in mundo sunt per umbras gnomonis inventae quibusque rationibus dilatentur aut contrahantur explicabo* (Eng. trans. M. H. Morris).

[15] Vitruvius: *De Architectura* IX, Praefactio, 3: *Cum ergo tanta munera ab scriptorum prudentia privatim publiceque fuerint hominibus praeparata, non solum arbitror palmas et coronas his tribui oportere, sed etiam decerni triumphos et inter deorum sedes eos dedicando iudicari* (Eng. trans. M. H. Morris).

Sun between Venus and Mars) held by Posidonius. As the Earth is concerned, although its sphericity is not explicitly claimed, the agreement with Posidonius is evident:

> The word "universe" means the general assemblage of all nature, and it also means the heaven that is made up of the constellations and the courses of the stars. The heaven revolves steadily round earth and sea on the pivots at the ends of its axis. The architect at these points was the power of Nature, and she put the pivots there, to be, as it were, centres, one of them above the earth and sea at the very top of the firmament and even beyond the stars composing the Great Bear, the other on the opposite side under the earth in the regions of the south. Round these pivots (termed in Greek *πολοι* as centres, like those of a turning lathe, she formed the circles in which the heaven passes on its everlasting way. In the midst thereof, the earth and sea naturally occupy the central point. It follows from this natural arrangement that the central point in the north is high above the earth, while on the south, the region below, it is beneath the earth and consequently hidden by it. Furthermore, across the middle, and obliquely inclined to the south, there is a broad circular belt composed of the twelve signs, whose stars, arranged in twelve equivalent divisions, represent each a shape which nature has depicted. And so with the firmament and the other constellations, they move round the earth and sea in glittering array, completing their orbits according to the spherical shape of the heaven.[16]

In Chaps. 7 and 8 of the same book, he describes how to construct sundials as well as a water clock. We have already quoted in Sect. 1.7 the information he provides regarding the "scaphe".

In carrying on our research of the current opinion on the nature of Earth in the Roman intellectual world, we now refer to Seneca, born about seventy years after Vitruvius. Seneca (4 BC–55 AD), born in Spain, held important official positions, was tutor to Nero and at the end was induced by him to commit suicide, being under indictment for having taken part in a conspiracy against him. Among his works, we are interested in *Naturales Quaestiones*, written, as it seems, between 62 and 63 AD, after his retreat from public life. This work is dedicated to his nephew Lucius, as is the *Ad Lucium Epistolae morales*, and resumed the studies to which he had attended in his youth. The source from which he drew the most information is undoubtedly a meteorological work of Posidonius. In the *Naturales Quaestiones*, he generally accepts Posidonious' opinion, not as a repeater, but as an unbiased researcher who recognizes as correct a truth enounced before him. The phenomena

---

[16]Vitruvius: *De Architectura*, IX, 1, 2–3: *mundus autem est omnium naturae rerum conceptio summa caelumque sideribus et stellarum cursibus conforrnatum. id volvitur continenter circum terram atque mare per axis cardines extremos. namque in his locis naturalis potestas ita architectata est conlocavitque cardines tamquam centra, unum a terra et mari in summo mundo ac post ipsas stellas septentrionum, alterum trans centra sub terra in meridianis partibus, ibique circum eos cardines orbiculos circum centra uti in torno perfecit, qui graece , πόλοι nominantur, per quos pervolitat sempiterno caelum. ita media terra cum mari centri loco naturaliter est conlocata. his natura dispositis ita uti septentrionali parte a terra excelsius habeat altitudinem centrum, in meridiana autem parte in inferioribus locis subiectum a terra obscuretur, tunc etiam per medium transversa et inclinata in meridiem circuli lata zona XII signis est conformata, quorum species stellis dispositis XII partibus peraequatis exprimit depictam ab natura figurationem. itaque lucentia cum mundo reliquisque sideribus ornatu circum terram mareque pervolantia cursus perficiunt ad caeli rotunditatem* (Eng. trans. Morgan Hicky Morgan).

of which he treats are mainly meteorological manifestations, such as winds, rain, hail, snow, comets, rainbows, and what he regards as allied subjects: earthquakes, springs, and rivers.

Like Ptolemy, Seneca believed in the celestial signs, that is, in astrology. The *Naturales Quaestiones* were much quoted in the Middle Ages and Adelard of Bath even copied their title.

We can now quote the passage of the *Naturales Quaestiones* which responds to our purpose:

> But contrariwise, certain persons assert that mountain peaks ought to be warmer in the degree in which they are nearer the sun. Such people seem to me, however, to be astray in supposing that the Apennines and the Alps and other mountains famed for their exceeding height are so greatly elevated that their size should enable them to feel in any special way the sun's proximity. No doubt those are lofty heights so long as the standard of comparison is ourselves. But when one regards the size of the universe, the lowness of them all becomes evident. Compared with one another, mountains are surpassed or surpass in height. But nothing on earth is elevated so high that even the greatest of objects should be any appreciable portion in comparison with the whole universe. Were this not so, we should not be in the habit of saying that the whole earth is a ball. The distinctive mark of a ball is a certain uniform rotundity, much the same as the uniformity seen in a football or cricket ball. The seams and chinks constitute no great objection to the ball being described as symmetrical on all sides. As in a playing ball, those spaces do not in any way prevent the appearance of roundness, no more, in the earth at large regarded as a sphere, do lofty mountains, whose height is lost in a comparison with the whole world. A person who says that a higher mountain ought to be warmer from receiving the sun's rays at a shorter distance, may just as well say that a taller man should be heated sooner than a dwarf, and his head sooner than his feet![17]

Therefore, we can see that Seneca was sure of the sphericity of the Earth and was ironical about the convictions of his opponents.

The *Chorographia* of Pomponius Mela, which we have already discussed, had been written about twenty years before the *Naturales Quaestiones*. Therefore, it seems to us (and below we shall find confirmation in Pliny), that during almost all the first century the current opinion was in favour of the sphericity of the Earth.

A generation after Seneca, we meet Pliny the Elder (Gaius Plinius Secundus, 23–79 AD), so called in order to distinguish him from his homonymous nephew

---

[17]Pliny: *Naturales Quaestiones*, IV, 11, 1–4: *Contra quidam aiunt cacumina montium hoc calidiora esse debere, quo propiora soli sunt: qui mihi uidentur errare, quod Apenninum et Alpes et alios notos ob eximiam altitudinem montes in tantum putant crescere, ut illorum magnitudo sentire solis uiciniam possiti excelsa sunt ista, quamdiu nobis comparantur; at uero, ubi ad uniuersum respexeris, manifesta est omnium humilitas. Inter se uincuntur et uincunt; ceterum in tantum nihil attollitur, ut in collatione totius ulla sit uel maximis portio: quod nisi esset, non diceremus totum orbem terrarum pilam esse. Pilae proprietas est cum aequalitate quadam rotunditas, aequalitatem autem hanc accipe quam uides in lusoria pila: non multum illi commissurae et rimae [earum] nocent quo minus par sibi ab omni parte dicatur. Quomodo in hac pila nihil illa interualla officiunt ad speciem rotundi, sic ne in uniuerso quidem orbe terrarum editi montes, quorum altitudo totius mundi collatione consumitur. Qui dicit altiorem montem, quia solem propius excipiat, magis calere debere, idem dicere potest longiorem hominem citius quam pusillum debere calefieri et caput citius quam pedes* (Eng. trans. John Clarke).

(obviously, Pliny the Younger, 61–112 AD). Pliny, born in Como, held important positions in the administration of the empire and was a close adviser of the emperor Vespasian. He died in 79 AD, a victim of the famous eruption of Vesuvius, when he was in command of the fleet at Misenum on the bay of Naples. His monumental work, *Naturalis Historia* (dedicated to the emperor) was published in 77 AD, two years before the author's death. It consists of 37 books, the first of which is devoted to explaining the content of the following 36 books and to exhibit a long list of the authors whose works he has consulted. The remaining books deal with cosmography (II), geography (III–VI), anthropology (VII); the next twenty five are dedicated to the animal kingdom (VIII–XI), to the plant kingdom (XII–XIX) and to the utility that the man can obtain from the animals and from the plants, i.e., from medical botany (XX–XXVII) and from medical zoology (XXVIII–XXXII); the last five books are devoted to mineralogy especially considered with regard to the usages in human life and in the arts; an invaluable and succinct history of the artists and their works closes the *Naturalis Historia*.

We have gone into the description of the content of the 37 books to stress the care with which Pliny had gathered information on every aspect of the natural sciences. Luckily, this work, despite its length and the countless summaries of it that were written, has survived in its entirety, and thanks to this can supply scholars with valuable information on many works that are now lost. Pliny selects and reports whole pieces of various works, whether directly, or second-hand. In fact, many times he quotes a Greek work, but what he reports is a version of it due to a Latin author, whose name he omits. Several scholars have verified how an original passage of a Greek work turns out to be altered through the subsequent Latin translations. As often as not Pliny gives the impression of working with scissors and paste and sometimes contradicts himself, expressing conflicting opinions on the same question in different places of his work.

Obviously, we are interested in what Pliny says in the second book about cosmography. His conception of the world derives from Posidonius, at least for the "formal" part. Instead, the religious vehemence on the eternity of the world, and on the impossibility for the man of understanding the world, etc., is his own:

> The world and this—whatever other name men have chosen to designate the sky whose vaulted roof encircles the universe, is fitly believed to be a deity, eternal, immeasurable, a being that never began to exist and never will perish. What is outside it does not concern men to explore and is not within the grasp of the human mind to guess. It is sacred, eternal, immeasurable, wholly within the whole, nay rather itself the whole, finite and resembling the infinite, certain of all things and resembling the uncertain, holding in its embrace all things that are without and within, at once the work of nature and nature herself.[18]

[18]Pliny: *Naturalis Historia*, II, I, 1–2: *Mundum et hoc—quocumque nomine alio caelum appellare libuit cuius circumflexu teguntur cuncta, numen esse credi par est, aeternum. inmen-sum, neque genitum neque interiturum uniquam. huius extera indagare nec interest hominum nec capit humanae coniectura mentis. sacer est, ae-ternus. inmensus, totus in toto, immo vero ipse totum, finitus et infinito similis,omnium rerum certus et similis incerto, extra intra cuncta complexus in se, idemque rerum naturae opus et rerum ipsa natura* (Eng. trans. H. Rackham).

He continues by calling mad those who want to evaluate the size of the world and those, such as Leucippus and Democritus, who, by anticipating Giordano Bruno, speak up for the plurality of worlds. In this he manifests a kind of religious wrath:

> That certain persons have studied, and have dared to publish, its dimensions, is mere madness; and again that others, taking or receiving occasion from the former, have taught the existence of a countless number of worlds, involving the belief in as many systems of nature, or, if a single nature embraces all the worlds, nevertheless the same number of suns, moons and other unmeasurable and innumerable heavenly bodies, as already in a single world; just as if owing to our craving for some end the same problem would not always encounter us at the termination of this process of thought, or as if, assuming it possible to attribute this infinity of nature to the artificer of the universe, that same property would not be easier to understand in a single world, especially one that is so vast a structure. It is madness, downright madness, to go out of that world, and to investigate what lies outside it just as if the whole of what is within it were already clearly known; as though, forsooth, the measure of anything could be taken by him that knows not the measure of himself, or as if the mind of man could see things that the world itself does not contain.[19]

As regards the spherical shape of the world, Pliny says that this is already confirmed by unanimity (we shall see further on that this consent on the sphericity of the world will clash on some problems):

> Its shape has the rounded appearance of a perfect sphere. This is shown first of all by the name of "orb" which is bestowed upon it by the general consent of mankind. It is also shown by the evidence of the facts: not only does such a figure in all its parts converge upon itself; not only must it sustain itself, enclosing and holding itself together without the need of any fastenings, and without experiencing an end or a beginning at any part of itself; not only is that shape the one best fitted for the motion with which, as will shortly appear, it must repeatedly revolve, but our eyesight also confirms this belief, because the firmament presents the aspect of a concave hemisphere equidistant in every direction, which would be impossible in the case of any other figure.[20]

[19]Pliny: *Naturalis Historia*, II, I, 3–4: *Furor est mensuram eius animo quosdam agitasse atque prodere ausos, alios rursus occasione hinc consumpta aut his data innumerabiles tradidisse mundos, ut totidem rerum naturas credi oporteret aut, si una omnes incubaret, totidem tamen soles totidemque lunas et cetera etiam in uno et inmensa et innumerabilia sidera, quasi non eaedem quaestiones semper in termino cogitationi sint occursurae desiderio finis alicuius aut, si haec infinitas naturae omnium artifici possit adsignari, non idem illud in uno facilius sit intellegi, tanto praesertìm opere. Furor est profecto, furor egredi ex eo et, tamquam interna eius cuncta piane iam nota sint, ita scrutati extera, quasi vero mensuram ullius rei possit agere qui sui nesciat, aut mens hominis videre quae mundus ipse non capiat* (Eng. trans. H. Rackham).

[20]Pliny: *Naturalis Historia*, II, II. 5: *Formam eius in speciem orbis absoluti globatam esse nomen in primis et consensus in eo mortalium orbem appellantium, sed et argumenta rerum docent. non solum quia talis figura omnibus sui partibus vergit in sese ac sibi ipsa toleranda est seque includit et continet nullarum egens compagium nec finem aut initium ullis sui partibus sentiens, nec quia ad motum, quo subinde verti mox adparebit, talis aptissima est, sed oculorum quoque probatione, quod convexus mediusque quacumque cernatur, cum id accidere in alia non possit figura* (Eng. trans. H. Rackham).

Pliny again invokes the consensus, on speaking of the sphericity of the Earth:

> But her shape is the first fact about which men's judgement agrees. We do undoubtedly speak of the earth's sphere, and admit that the globe is shut in between poles. Nor yet in fact do all these lofty mountains and widely spreading plains comprise the outline of a perfect sphere, but a figure whose circuit would produce a perfect sphere if the ends of all the lines were enclosed in a circumference. This is the consequence of the very nature of things, it is not due to the same causes as those we have adduced in the case of the heaven; for in the heaven the convex hollow converges on itself and from all sides rests upon its pivot, the earth, whereas the earth being a solid dense mass rises like an object swelling, and expands outward. The world converges to its centre, whereas the earth radiates outward from its centre, the ceaseless revolution of the world around her forcing her immense globe into the shape of a sphere.[21]

The consensus, however, falls apart when, once the sphericity of the Earth is accepted, one bumps into the problem of the Antipodes. In this case, the opinion of the learned persons is quite far apart from that of the man in the street. We have already faced this problem and we shall be up against it for centuries. Now, let us listen to Pliny:

> Here there is a mighty battle between learning on one side and the common herd on the other: the theory being that human beings are distributed all round the earth and stand with their feet pointing towards each other, and that the top of the sky is alike for them all and the earth trodden under foot at the centre in the same way from any direction, while ordinary people enquire why the persons on the opposite side don't fall off—just as if it were not reasonable that the people on the other side wonder that we do not fall off. There is an intermediate theory that is acceptable even to the unlearned crowd—that the earth is of the shape of an irregular globe, resembling a pine cone, yet nevertheless is inhabited all round. But what is the good of this theory when there arises another marvel, that the earth herself hangs suspended and does not fall and carry us with it? As if forsooth there were any doubt about the force of breath, especially when shut up inside the world, or as if it were possible for the earth to fall when nature opposes, and denies it any place to fall to! For just as the sole abode of fires is in the element of fire, and of waters in water, and of breath in breath, so earth, barred out by all the other elements, has no place except in itself. Yet it is surprising that with this vast level expanse of sea and plains the resulting formation is a globe.[22]

---

[21]Pliny: *Naturalis Historia*, II, LXIV, 160: *Est autem figura prima de qua consensus iudicat. orbem certe dicimus terrae, globumque verticibus includi fatemur. neque enim absoluti orbis est forma in tanta montium excelsitate, tanta camporum planitie, sed cuius amplexus, si capita cunctarum liniarum conprehendantur ambitu, figuram absoluti orbis efficiat— id quod ipsa rerum natura cogit, non eisdem causis quas attulimus in caelo. namque in illo cava in se convexitas vergit et cardini suo, hoc est terrae, undique incumbit, haec ut solida ac conferta adsurgit intumescenti similis extraque protenditur. mundus in centrum vergit, at terra exit a centro, immensum eius globum in formam orbis adsidua circa eam mundi volubilitate cogente* (Eng. trans. H. Rackham).

[22]Pliny: *Naturalis Historia*, II, LXV, 161–162: *Ingens hic pugna litterarum contraque volgi: circumfundi terrae undique homines conversisque inter se pedibus stare, et cunctis similem esse caeli verticem, simili modo ex quacumque parte mediam terram calcari, illo quaerente, cur non decidant contra siti, tamquam non ratio praesto sit ut nos non decidere mirentur illi. intervenit sententia quamvis indocili probabilis turbae, inaequali globo, ut si sit figura pineae nucis, nihilominus terram undique incoli. Sed quid hoc refert alio miraculo exoriente, pendere ipsam ac*

However, the author does not stop here and quotes the evidence provided by the Greek philosophers, starting from Dicearchus, and tries to demonstrate that water also arranges itself in conformity of a spherical surface. Finally, at the end of the book, contradicting himself (previously he had condemned those who wanted to measure the world), he seems to feel admiration for the measure of the circumference of the Earth performed by Eratosthenes (by quoting the famous figure of 252,000 stades). Not satisfied of this, he also reports the fantastic story of a certain Dyonisodorus who had left, inside his grave, a signed letter in which he told of having gone down to the centre of the Earth and of having found that the distance was of 42,000 stades; on the basis of this figure the geometricians could have calculated that the terrestrial circumference was of 252,000 stades:

> These are the facts that I consider worth recording in regard to the earth's length and breadth. Its total circumference was given by Eratosthenes (an expert in every refinement of learning, but on this point assuredly an outstanding authority—I notice that he is universally accepted) as 252,000 stades, a measurement that by Roman reckoning makes 31,500 miles—an audacious venture, but achieved by such subtle reasoning that one is ashamed to be sceptical. Hipparchus, who in his refutation of Eratosthenes and also in all the rest of his researches is remarkable, adds a little less than 26,000 stades. Dionysodorus (for I will not withhold this outstanding instance of Greek folly) has a different creed. He belonged to Melos, and was a celebrated geometrician; his old age came to its term in his native place; his female relations who were his heirs escorted his obsequies. It is said that while these women on the following days were carrying out the due rites they found in the tomb a letter signed with his name and addressed to those on earth, which stated that he had passed from his tomb to the bottom of the earth and that it was a distance of 42,000 stades. Geometricians were forthcoming who construed this to mean that the letter had been sent from the centre of the earth's globe, which was the longest space downward from the surface and was also the centre of the sphere. From this the calculation followed that led them to pronounce the circumference of the globe to be 252,000 stades.[23]

non cadere nobiscum?—ceu spiritus vis. mundo praesertim inclusi, dubia sit, aut possit cadere natura repugnante et quo cadat negante! nam sicut ignium sedes non est nisi in ignibus, aquarum nisi in aquis spiritus nisi in spiritu, sic terrae arcentibus cunctis nisi in se locus non est. globum tamen effici mirum est in tanta planitie maris camporumque (Eng. trans. H. Rackham).

[23]Pliny: *Naturalis Historia*, II, CXII, 247–248: *De longitudine ac latitudine haec sunt quae digna memoratu putem. universum autem circuitum Eratosthenes (in omnium quidem litterarum subtilitate set in hac utique praeter ceteros solers, quem cunctis probari video) CCLII milium stadiorum prodidit, quae mensura Romana conputatione efficit trecentiens quindeciens centena milia passuum, improbum ausum, verum ita subtili argumentatione comprehensum ut pudeat non credere. Hipparchus et in coarguendo eo et in reliqua omni diligentia mirus, adicit stadiorum paulo minus XXVI. Alia Dionysodoro fides (neque enim subtraham exemplum vanitatis Graecae maximum). Melius hic fuit geometricae scientia nobilis; senecta diem obiit in patria, funus duxere ei propinquae ad quas pertinebat hereditas. hae cum secutis diebus insta peragerent, invenisse dicuntur in sepulcro epistulam Dionysodori nomine ad superos scriptam: pervenisse eum a sepulcro ad infimam terram, esse eam stadiorum XLII. Nec defuere geometrae qui interpretarentur significare epistulam a medio terrarum orbe missam quod deorsum ab summo longissimum esset spatium et idem pilae medium, ex quo consecuta computatio est circuitum esse CCLII stadiorum pronuntiarentur* (Eng. trans. H. Rackham).

Pliny's work, as we have seen above, received much attention, both immediate and posthumous, and thus one would expect that what asserted by him with regard to cosmography and geography was largely shared, at least by the intellectuals. Instead, about twenty years after Pliny's death, Tacitus (Cornelius Tacitus, 56–117 AD) seems to still think like the Ionic School, that is, he still imagines the Earth as disc-shaped. Let us outline the question.

Tacitus, maybe native to Belgian Gaul, besides having been a great historian (his works date back to the last year of the first century and to the first decade of the second) also held important positions in the administration of the empire. He was a close friend of Pliny the Younger and had applied to the younger man to learn how Pliny's died (as we know from a letter of Pliny the Younger; see *Epistulae*, VI, 16), so one might conjecture that he knew the *Naturalis Historia* or, at least, he was well-informed on the question concerning the shape of the Earth. But instead, here is what he says in his work *The Life and Death of Julius Agricola* (a biography of his father-in-law, maybe written in 97–98 AD):

> Their sky is obscured by continual rain and cloud. Severity of cold is unknown. The days exceed in length those of our part of the world; the nights are bright, and in the extreme north so short that between sunlight and dawn you can perceive but a slight distinction. It is said that, if mere are no clouds in the way, the splendour of the sun can be seen throughout the night, and that he does not rise and set, but only crosses the heavens. The truth is, that the low shadow thrown from the flat extremities of the earth's surface does not raise the darkness to any height, and the night thus fails to reach the sky and stars.[24]

The English translation of this passage (of which the Latin text is given in the footnote) interprets the original ascribing, obviously, to Tacitus the opinion that the Earth is disc-shaped. We too think that this is a correct interpretation of the Latin text. This circumstance, however, astounded several translators, who were perhaps inclined to ascribe the origin of a mistake to the well-known Tacitean *brevitas*. Many scholars of different provenance have expressed themselves, some attributing to Tacitus the conception of a spherical Earth, others that of a disc-shaped Earth. The question is mentioned by the Italian Hellenist Pietro Janni who reports three Italian translations of different interpretations, and also provides a bibliography on the matter.[25]

Actually, the question amazes us since, in accordance with the habit of mind of us moderns, we are driven to think that at that time the whole body of Roman learned people should have absorbed and accepted that the Earth had a spherical shape. Instead, our astonishment must be also extended to the Greek learned world of the same age. Also in this case, we have to do with a historian, a great historian,

---

[24]Tacitus: *De vita et moribus Iulii Agricolae*, 12: [5] *Caelum crebris imbribus ac nebulis foedum; asperitas frigorum abest.* [6] *Dierum spatia ultra nostri orbis mensuram; nox clara et extrema Britanniae parte brevis, ut finem atque initium lucis exiguo discrimine internoscas.* [7] *Quod si nubes non offlciant, aspici per noctem solis fulgorem, nec Decidere et exsurgere, sed transire adfirmant.* [8] *Scilicet extrema et plana terrarum humili umbra non erigunt tenebras, infraque caelum et sidera nox cadit* (Eng. trans. Alfred John Church and William Jackson Brodribb).

[25]Pietro Janni: *Miti e falsi miti* (Dedalo, 2004), p. 149.

even if of a type completely different from Tacitus. This is Plutarch of Chaeronea (45–120 AD), renowned over the centuries for his work *Parallel Lives* (printed and disseminated throughout the Latin West starting from the end of the fifteenth century, and taken as a fundamental text for the "*cursus studiorum*" of the learned classes until the nineteenth century).

In addition to *Parallel Lives*, Plutarch also wrote a series of small works which were collected in a single corpus (by Maximus Planudes in 1296) under the title *Moralia*. It is one of these works, entitled *On the face in the Moon* (*De facie quae in orbe lunae apparet*) that we want to speak of (already quoted by us in 1.7). This is a brief work in the form of a dialogue (the interlocutors are eight: five Greeks, one Oriental, and two Romans, Lucius and Sulla) whose subject is apparently the structure, the constitution and the appearance of the Moon. In the manuscript tradition the work presents a lacuna in the beginning[26] and in the discussion there are references to an earlier lecture given by "our comrade" which is recapitulated by Lamprias, Plutarch's brother. Actually, the work is a digression which also concerns cosmography and geography.

Plutarch is undoubtedly the most brilliant intellectual in the Hellenistic world of his age, thus, one would expect him to be familiar with the progress of the science of the time and, in consequence, to provide an accurate and an up-to-date discussion. Instead, his points of reference span a wide interval (ranging from Empedocles to his contemporaries) and he seems to ignore what had been a commonplace since the age of Aristotle and the Stoics. Let us look at a passage which confirms what above said:

> …we must not listen to philosophers, if they claim to meet paradoxes with paradoxes, and controvert surprising doctrines by inventing others still more strange and surprising, as these people do with their idea of motion towards the centre. What absurdity is there that this does not imply? Does it not mean that the earth is a sphere, though it contains such enormous depths and heights and irregularities? That people dwell at our antipodes, like wood-worms or lizards, clinging to the earth with their lower limbs upwards? That we ourselves do not remain perpendicular as we walk, but remain at an angle and sway like drunken men?[27]

Apuleius from Madaurus, born about eighty years after Plutarch (126–about 180 AD), a Latin writer of the period of the Antonines, can be considered the last important Latin writer of the period of the Latin pagan literature. He is mentioned in the histories of Latin literature for his novel *The Golden Ass* or *The Metamorphoses*, but the work of his that we are interested in is *De Mundo*. This writing is a translation, with literary pretences, of a work ascribed to Aristotle. This attribution has been disputed for years, and only some decades ago Giovanni Reale, a scholar of Aristotle's works, has confirmed the attribution.[28] We have just said

[26]For this see H. Martin, Jr.: "Plutarch's *De facie*: the Recapitulation and the Lost Beginning", *Greek, Roman and Byzantine Studies* 15 (1974): 73–88.

[27]Plutarch: *On the Face in the Moon*, op. cit., 7 (Eng. trans. T. L. Heath, *Greek Astronomy*, op. cit., pp 171–172).

[28]See G. Reale (ed.): *Trattato sul Cosmo per Alessandro* (Naples, 1974).

that the work of Apuleius is not a translation in the literal meaning of the word, but has literary pretences, as if the original work had been a pretext for a literary work. This is, at least, the opinion of scholars. We are only interested in remarking that, even if enriched with literary stylistic elements, the substance of Aristotelian cosmography in any case remains: the Earth is considered to have a spherical shape and the axis of the cosmos, which rotates, holding the orb of the Earth in the centre — "…*orbem terrae in medietate constituens*".[29]

About a century lapsed between the death of Apuleius and Diocletian's coming into power (284 AD). The Roman prosaists of this period (at least in what is extant) do not deal with subjects comprising cosmography or geography. We shall see in the next section how the birth of a Christian literature will once again raise the problem of the shape and structure of the Earth.

2.3 Late Antiquity. The Decline of the Roman Empire

Conventionally, it has been established that the end of the period called the Late Antiquity (and then the beginning of the Middle Ages) coincides with the fall of the Western Roman Empire (476 AD, the deposition of Romulus Augustulus by Odoacer). The period of emperor Diocletianus (284–305 AD) witnessed the last persecutions of the Christians took place, but it also witnessed the growth in conversions to the Christian faith of learned Roman citizens. One convert of particular relevance for the future history of the Church, and also for the subject of our interest, was Lucius Caecilius Firmianus Lactantius.

Together with Tertullian and Minucius Felix, Lactantius is one of the protagonists of the group called of the Apologists, that is, those writers who individually devoted themselves to defending orthodoxy and opposing heresy and paganism. Here we are obviously referring to the Western Apologists. Whereas the Eastern Apologists tried to establish a continuity with the Greek philosophy and presented the Christian doctrine as the true philosophy that Christ's revelation carried to its final accomplishment, the Western Apologists aimed to claim the originality of the Christian revelation with respect to pagan wisdom and to found it on the pratical and immediate nature of the faith, rather than on the inquiry.

This character of the Latin apologies is already evident in its most important representative, Tertullian (ca 155–220 AD), the first Latin Apologist and also the first Latin Christian writer. He was a lawyer and fought, with his forensic skill and legal procedures, both the pagans and the heretical Christians. The other Apologist was Minucius Felix (II–III century AD) who, although he too was a lawyer, rather than resorting to forensic procedures, resorted to the philosophical dialogue (following the example of Plato and Cicero).

---

[29]Apuleius: *De Mundo*, 230.

2.3.1 Lactantius

As we have anticipated, our interest is in Lactantius (250–ca 326 AD), who lived during a crucial change of the official politics towards the Christian faith, marked by the edict of Milan, promulgated in 312 AD by the emperor Constantine. As he himself narrates, he was of African and pagan origin. As a young man, he stood out for his oratorical skills, and because of that was sent by Diocletian to Nicomedia to teach Latin rethoric. He integrated that activity with the writing of school treatises (all now lost).

In 303 was promulgated the great persecution against the Christians by Diocletian, but Lactantius remained professor of rethoric in Nicomedia until Galerius (in 306) ordered the closure of the schools, causing Lactantius to lose his chair. His conversion to Christian faith, which occurred in silence, may date back to this time. After a period of relative isolation, devoted to the writing of his most important works (*De Opificio Dei* and *Divinae Institutiones*), in 316–317 he was entrusted by the emperor Constantine with the task of educating his son Crispus. That reminds the historic precedent of Aristotle and Alexander (although Crispus had a different fate). In the last period of his life, Lactantius ended up in the important official position of tutor to the emperor's son. We must insist on this last point since we think that, precisely for this reason, the passages of *Divinae Institutiones* that we will quote must be considered as having received (in those years) particular attention.

We mentioned that Lactantius had a sound background in classic culture, that is, he knew both the Greek philosophers and the Latin writers, therefore what he wrote was not due to ignorance (even if St. Jerome imputed to him some slips precisely regarding Christian theology). Let us begin with his invective against the philosophers (obviously the Greek philosophers and their Roman divulgers):

> For they are always deceived in the same manner. For when they have assumed anything false in the commencement of their investigations, led by the resemblance of the truth, they necessarily fall into those things which are its consequences. Thus they fall into many ridiculous things; because those things which are in agreement with false things, must themselves be false. But since they placed confidence in the first, they do not consider the character of those things which follow, but defend them in every way; whereas they ought to judge from those which follow, whether the first are true or false.[30]

Now let us look at his argumentations:

> What course of argument, therefore, led them to the idea of the antipodes? They saw the courses of the stars travelling towards the west; they saw that the sun and the moon always set towards the same quarter, and rise from the same. But since they did not perceive what contrivance regulated their courses, nor how they returned from the west to the east, but supposed that the heaven itself sloped downwards in every direction, which appearance it must present on account

[30]Lactantius: *Divinae Institutiones*, III, 24: *Nam semper eodem modo falluntur. Cum enim falsum aliquid in principio sumpserint, veri similitudine inducti, necesse est eos in ea, quae consequuntur, incurrere. Sic incidunt in multa ridicula; quia necesse est falsa esse, quae rebus falsis congruunt. Cum autem primis habuerint fidem, qualia sint ea, quae sequuntur, non circumspiciunt, sed defendunt omni modo; cum debeant prima illa, utrumne vera sint, an falsa, ex consequentibus judicare* (Eng. trans. William Fletcher).

of its immense breadth, they thought that the world is round like a ball, and they fancied that the heaven revolves in accordance with the motion of the heavenly bodies; and thus that the stars and sun, when they have set, by the very rapidity of the motion of the world are borne back to the east. Therefore they both constructed brazen orbs, as though after the figure of the world, and engraved upon them certain monstrous images, which they said were constellations. It followed, therefore, from this rotundity of the heaven, that the earth was enclosed in the midst of its curved surface. But if this were so, the earth also itself must be like a globe; for that could not possibly be anything but round, which was held enclosed by that which was round. But if the earth also were round, it must necessarily happen that it should present the same appearance to all parts of the heaven; that is, that it should raise aloft mountains, extend plains, and have level seas. And if this were so, that last consequence also followed, that there would be no part of the earth uninhabited by men and the other animals. Thus the rotundity of the earth leads, in addition, to the invention of those suspended antipodes. But if you inquire from those who defend these marvellous fictions, why all things do not fall into that lower part of the heaven, they reply that such is the nature of things, that heavy bodies are borne to the middle, and that they are all joined together towards the middle, as we see spokes in a wheel; but that the bodies which are light, as mist, smoke, and fire, are borne away from the middle, so as to seek the heaven. I am at a loss what to say respecting those who, when they have once erred, consistently persevere in their folly, and defend one vain thing by another; but that I sometimes imagine that they either discuss philosophy for the sake of a jest, or purposely and knowingly undertake to defend falsehoods, as if to exercise or display their talents on false subjects. But I should be able to prove by many arguments that it is impossible for the heaven to be lower than the earth, were it not that this book must now be concluded, and that some things still remain, which are more necessary for the present work. And since it is not the work of a single book to run over the errors of each individually, let it be sufficient to have enumerated a few, from which the nature of the others may be understood.[31]

[31]Lactantius: *Divinae Institutiones*, III, 24: *Quae igitur illos ad Antipodas ratio perduxit? Videbant siderum cursus in occasum meantium; solem atque lunam in eamdem partem semper occidere, atque oriri semper ab eadem. Cum autem non perspicerent, quae machinatio cursus eorum temperaret, nec quomodo ab occasu ad orientem remearent, coelum autem ipsum in omnes partes putarent esse devexum, quod sic videri, propter immensam latitudinem necesse est: existimaverunt, rotundum esse mundum sicut pilam, et ex motu siderum opinati sunt coelum volvi, sic astra solemque, cum occiderint, volubilitate ipsa mundi ad ortum referri. Itaque et aereos orbes fabricati sunt, quasi ad figuram mundi, eosque caelarunt portentosis quibusdam simulacris, quae astra esse dicerent. Hanc igitur coeli rotunditatem illud sequebatur, ut terra in medio sinu ejus esset inclusa. Quod si ita esset, etiam ipsam terram globo similem; neque enim fieri posset, ut non esset rotundum, quod rotundo conclusum teneretur. Si autem rotunda etiam terra esset, necesse esse, ut in omnes coeli partes eamdem faciem gerat, id est montes erigat, campos tendat, maria consternat. Quod si esset, etiam sequebatur illud extremum, ut nulla sit pars terrae, quae non ab hominibus caeterisque animalibus incolatur. Sic pendulos istos Antipodas coeli rotunditas adinvenit. Quod si quaeras ab iis, qui haec portenta defendunt, quomodo non cadunt omnia in inferiorem illam coeli partem; respondent, hanc rerum esse naturam, ut pondera in medium ferantur, et ad medium connexa sint omnia, sicut radios videmus in rota; quae autem levia sunt, ut nebula, fumus, ignis, a medio deferantur, ut coelum petant. Quid dicam de iis nescio, qui, cum semel aberraverint, constanter in stultitia perseverant, et vanis vana defendunt; nisi quod eos interdum puto, aut joci causa philosophari, aut prudentes et scios mendacia defendenda suscipere, quasi ut ingenia sua in malis rebus exerceant, vel ostendant. At ego multis argumentis probare possem, nullo modo fieri posse, ut coelum terra sit inferius, nisi et liber jam concludendus esset, et adhuc aliqua restarent, quae magis sunt praesenti operi necessaria. Et quoniam singulorum errores percurrere non est unius libri opus, satis sit pauca enumerasse, ex quibus possit qualia sint caetera intelligi* (Eng. trans. William Fletcher).

Previously, he had said:

How much better, therefore, is it, leaving vain and insensible objects, to turn our eyes in that direction where is the seat and dwelling-place of the true God; who suspended the earth on a firm foundation, who bespangled the heaven with shining stars; who lighted up the sun, the most bright and matchless light for the affairs of men, in proof of His own single majesty; who girded the earth with seas, and ordered the rivers to flow with perpetual course!

He also commanded the plains to extend themselves, the valleys to sink down, the woods to be covered with foliage, the stony mountains to rise.

All these things truly were not the work of Jupiter, who was born seventeen hundred years ago; but of the same, that framer of all things, the origin of a better world, who is called God, whose beginning cannot be comprehended, and ought not to be made the subject of inquiry. It is sufficient for man, to his full and perfect wisdom, if he understands the existence of God: the force and sum of which understanding is this, that he look up to and honour the common Parent of the human race, and the Maker of wonderful things. Whence some persons of dull and obtuse mind adore as gods the elements, which are both created objects and are void of sensibility; who, when they admired the works of God, that is, the heaven with its various lights, the earth with its plains and mountains, the seas with their rivers and lakes and fountains, struck with admiration of these things, and forgetting the Maker Himself, whom they were unable to see, began to adore and worship His works. Nor were they able at all to understand how much greater and more wonderful He is, who made these things out of nothing. And when they see that these things, in obedience to divine laws, by a perpetual necessity are subservient to the uses and interests of men, they nevertheless regard them as gods, being ungrateful towards the divine bounty, so that they preferred their own works to their most indulgent God and Father. But what wonder is it if uncivilized or ignorant men err, since even philosophers of the Stoic sect are of the same opinion, so as to judge that all the heavenly bodies which have motion are to be reckoned in the number of gods; ... And we prove that you, o philosophers, are not only unlearned and impious, but also blind, foolish, and senseless, who have surpassed in shallowness the ignorance of the uneducated.[32]

[32]Lactantius: *Divinae Institutiones*, II, 5: *Quanto igitur rectius est, omissis insensibilibus et vanis, oculos eo tendere, ubi sedes, ubi habitatio est Dei veri; qui terram stabili firmitate suspendit; qui coelum distinxit astris fulgentibus; qui solem rebus humanis clarissimum, ac singulare lumen, in argumentum suae unicae majestatis accendit: terris autem maria circumfudit, flumina sempiterno lapsu fluere praecepit. Jussit et extendi campos, subsidere valles, Fronde tegi silvas, lapidosos surgere montes. ·uae utique omnia non Jupiter fecit, qui ante annos mille septingentos natus; sed idem: Ille opifex rerum, mundi melioris origo, qui vocatur Deus, cujus principium, quoniam non potest comprehendi, ne quaeri quidem debet. Satis est homini ad plenam perfectamque prudentiam, si Deum esse intelligat: cujus intelligentiae vis et summa haec est, ut suspiciat et honorificet communem parentem generis humani, et rerum mirabilium fabricatorem. Unde quidam hebetis obtusique cordis, elementa, quae et facta sunt et carent* sensu, *tamquam deos adorant. Qui cum Dei opera mirarentur, id est coelum cum variis luminibus, terram cum campis et montibus, maria cum fluminibus et stagnis et fontibus, earum rerum admiratione obstupefacti, et ipsius artificis obliti, quem videre non poterant, ejus opera venerari et colere coeperunt; nec umquam intelligere quiverunt, quanto major quantoque mirabilior sit, qui illa fecit ex nihilo. Quae cum videant divinis legibus obsequentia commodis atque usibus hominis perpetua necessitate famulari, tamen illa deos existimant esse; ingrati adversus beneficia divina, qui Deo et patri indulgentissimo sua sibi opera praetulerunt. Sed quid mirum, si aut barbari, aut imperiti homines errant? cum etiam philosophi Stoicae disciplinae in eadem sint opinione, ut omnia coelestia, quae moventur, in deorum numero habenda esse censeant; ... ac vos, o philosophi, non solum indoctos et impios, verum etiam caecos, ineptos delirosque probamus, qui ignorantiam imperitorum vanitate vicistis* (Eng. trans. William Fletcher).

This acrimony towards the "philosophers" was the outcome of the tangled situation in which the Apologists (and, after them, the Fathers of the Church) found themselves. When applying to learned persons they felt bound to recover the (Hellenistic-Roman) cultural tradition, but at the same time they were absolutely obliged to remain attached to what was written in the Bible.

The book of Genesis (if interpreted literally) was in patent contradiction with the results obtained by the science up to that time. One can say that henceforth in the Western Christian world the conflict begins between science and faith (a conflict which has never abated, although it has transformed over time).

Let us review the first verses of Genesis:

> [1]In the beginning when God created the heavens and the earth, [2]the earth was a formless void and darkness covered the face of the deep, while a wind from God swept over the face of the waters. [3]Then God said, "Let there be light"; and there was light. [4]And God saw that the light was good; and God separated the light from the darkness. [5]God called the light Day, and the darkness he called Night. And there was evening and there was morning, the first day.
>
> [6]And God said, "Let there be a dome in the midst of the waters, and let it separate the waters from the waters." [7]So God made the dome and separated the waters that were under the dome from the waters that were above the dome. And it was so. [8]God called the dome Sky. And there was evening and there was morning, the second day.
>
> [9]And God said, "Let the waters under the sky be gathered together into one place, and let the dry land appear." And it was so. [10]God called the dry land Earth, and the waters that were gathered together he called Seas. And God saw that it was good.[33]

The interpretation of the seventh verse (that of the supercelestial waters) got the Christian scientists into difficulty for a long time. Obviously, it has been always influenced in a determinant way by the stage of scientific knowledge achieved in each single age. What is evident, is that the verses of Genesis (if read literally) cannot in any way be reconciled with the two-sphere model of the Ptolemaic cosmology.

---

[33]Genesis 1, 1: *In principio creavit Deus caelum et terram / terra autem erat inanis et vacua et tenebrae super faciem abyssi et spiritus Dei ferebatur super aquas / dixitque Deus fiat lux et facta est lux / et vidit Deus lucem quod esset bona et divisit lucem ac tenebras / appellavitque lucem diem et tenebras noctem factumque est vespere et mane dies unus / dixit quoque Deus fiat firmamentum in medio aquarum et dividat aquas ab aquis / et fecit Deus firmamentum divisitque aquas quae erant sub firmamento ab his quae erant super firmamentum et factum est ita / vocavitque Deus firmamentum caelum et factum est vespere et mane dies secundus / dixit vero Deus congregentur aquae quae sub caelo sunt in locum unum et appareat arida factumque est ita / et vocavit Deus aridam terram congregationesque aquarum appellavit maria et vidit Deus quod esset bonum* (Eng. trans. New Revised Standard Version).

2.3.2 St. Ambrose

We shall now be interested in a figure completely different from Lactantius. Whereas the latter was an intellectual devoted to the apology of Christianity, Ambrose (339–397 AD) would pass from the state administration to that of the Church as bishop of Milan.

Ambrose was probably born in Trier into a noble consular Christian family. After his father's death withdrew to Rome with his mother and was educated and initiated to the highest imperial positions. While still young he became the governor of Liguria and Emilia with residence in Milan. In his capacity as governor he found himself having to quell a riot that erupted in the Milanese community after the death of the Arian bishop Auxentius. Ambrose, relying on the favour he inspired in the people, not only succeeded in calming heated feelings but was even acclaimed bishop. He accepted the office and henceforth he played a dominant role in the whole Western Church, in both the discipline and the liturgy.

From the doctrinal point of view, Ambrose was an autodidact but very soon he became an authority on the subject and wrote several works. We shall refer to the most renowned of his works, the *Hexaemeron* (in six books, one for each day of the creation), a collection of sermons preached during the Lent. Therefore, in this work his doctrine is called upon to glorify the creation of the cosmos. Let us look at the principles he enunciates:

> To such an extent have men's opinions varied that some, like Plato and his pupils, have established three principles for all things; that is, God, Idea, and Matter. The same philosophers hold that these principles are uncreated, incorruptible, and without a beginning. They maintain that God, acting not as a creator of matter but as a craftsman who reproduced a model, that is, an Idea, made the world out of matter. This matter, which they call ὕλη, is considered to have given the power of creation to all things. The world, too, they regard as incorruptible, not created or made. Still others hold opinions such as those which Aristotle considered worthy of being discussed with his pupils. These postulate two principles, matter and form, and along with these a third principle which is called "efficient", which Aristotle considered to be sufficient to bring effectively into existence what in his opinion should be initiated.
>
> What, therefore, is more absurd than to link the eternity of the work of creation with the eternity of God the omnipotent? Or to identify the creation itself with God so as to confer divine honors on the sky, the earth, and the sea? From this opinion there proceeds the belief that parts of the world are gods. Yet on the constitution of the world itself there is no small difference of opinion among philosophers.
>
> Pythagoras maintains that there is one world. Others say that the number of worlds is countless, as was stated by Democritus, whose treatment of the natural sciences has been granted the highest authority by the ancients. That the world always was and always will be is the claim of Aristotle. On the other hand, Plato ventures to assert that the world did not always exist, but that it will always exist. A great many writers, however, give us evidence from their works that they believe that the world did not always exist and that it will not exist forever.
>
> How is it possible to arrive at an estimate of the truth amid such warring opinions? Some, indeed, state that the world itself is God, inasmuch as they consider that a divine mind seems to be within it, while others maintain that God is in parts of the world; others still,

> that He is in both in which case it would be impossible to determine what is the appearance of God, or what is His number, position, life, or activity. If this evaluation of the world be followed, we have to understand God to be without sense, something which rotates, is round, is aflame, and impelled by certain movements something driven, not by its own force, but by something external to it.[34]

Regarding the Earth, he writes:

> On the nature and position of the earth there should be no need to enter into discussion at this point with respect to what is to come. It is sufficient for our information to state what the text of the Holy Scriptures establishes, namely, that, he hanged the earth upon nothing.
>
> What need is there to discuss whether the earth hangs in the air or rests on the water? From this would arise a controversy as to whether the nature of the air which is slight and yielding is such as to sustain a mass of earth; also, the question would arise, if the earth rested on the waters, would not the earth by its weight fall and sink into the waters? …
>
> Does not God clearly show that all things are established by His majesty, not by number, weight, and measures? For the creature has not given the law; rather, he accepts it or abides by that which has been accepted.[35]

We are reporting the opinions of the most important among the apologists and the Fathers of the Church (further on we shall see Augustine) on the structure of the cosmos and the nature of the Earth, not with a mere documentary intent, but

---

[34]*Hexaemeron*, I, 1: *Tantumne opinionis assumpisse homines, ut aliqui eorum tria principia constituerent omnium, Deum, et exemplar, et materiam, sicut Plato discipulique ejus; et ea incorrupta, et increata, ac sine initio esse asseverarent: Deumque non tanquam creatorem materiae, sed tanquam artificem ad exemplar, hoc est, ideam intendentem, fecisse mundum de materia, quam vocant hylen, quae gignendi causas rebus omnibus dedisse asseratur: ipsum quoque mundum incorruptum, nec creatum aut factum existimarent: alii quoque, ut Aristoteles cum suis disputandum putavit, duo principia ponerent, materiam et speciem, et tertium cum iis, quod operatorium dicitur, cui suppeteret competenter efficere, quod adoriundum putasset.*

Quid igitur tam inconveniens, quam ut aeternitatem operis cum Dei omnipotentis asternitate conjungerent, vel ipsum opus Deum esse dicerent; ut coelum, et terram, et mare divinis prosequerentur honoribus? Ex quo factum est, ut partes mundi deos esse crederent, quamvis de ipso mundo non mediocris inter eos quaestio sit.

Nam Pythagoras unum mundum asserit: alii innumerabiles dicunt esse mundos, ut scripsit Democritus, cui plurimum de physicis vetustas auctoritatis detulit. Ipsumque mundum semper fuisse et fore Aristoteles usurpat dicere: contra autem Plato non semper fuisse, sed semper fore praesumit astruere? plurimi vero nec fuisse semper, nec semper fore scriptis suis testficantur.

Inter has eorum dissensiones quae potest esse veri aestimatio? cum alii mundum ipsum Deum esse dicant, quod ei mens divina, ut putant, inesse videatur: alii partes ejus, alii utrumque: in quo nec quae figura sit deorum, nec qui numerus, nec qui locus, aut vita possit, aut cura comprehendi. Siquidem mundi aestimatione volubilem, rotundum, ardentem, quibusdam incitatum motibus, sine sensu *Deum conveniat intelligi, qui alieno, non suo motu feratur* (Eng. trans. John J. Savage).

[35]*Hexaemeron*, I, 6: *De terrae quoque vel qualitate vel positione tractarc nihil prodest ad spem futuri, cum satis sit ad scentiam quod Scripturarum divinarum series comprehendit, quia suspendit terram in nihilo. Quid nobis discutere utrum in aere pendeat, an super aquam, ut inde nascatur controversia, quomodo aeris natura tenuis et mollior molem possit sustentare terrenam? Aut quomodo, si super aquas non demergatur in aquam gravis ruina terrarum? ...*

Nonne evidenter ostendit Deus omnia majestate sua consistere in numero, pondere atque mensura? Neque enim creatura legem tribuit, sed accepit, et servat acceptam (Eng. trans. John J. Savage).

because we think that from them and with them has come into being the problem which still obsesses the Christian world: the conflict between science and faith. The figures we are talking about were not thinkers more or less in agreement among themselves and in conflict with the philosophers that had preceded them, but *maîtres à penser* capable of imposing (in the name of God and by the authority they wielded) their theories to the community of the faithful, even in the centuries to come. Secular thought, independent and unattached to the verses of Genesis, will be obliged to fight for centuries in order to emerge.

2.3.3 St. Augustine

We conclude now our inquiry into the opinions of the Apologists and of the Fathers of the Church, before the fall of the Roman Empire, with St. Augustine. Aurelius Augustinus was born in 374 AD in Tagaste (Numidia, in Roman Africa, now Algeria). His father Patrice was pagan and his mother Monica Christian, indeed very Christian. He carried out his university education with great success, first at Madaura and then at Carthage. He spent his early youth between his educational successes and the abuses of earthly pleasures (as he narrates in the *Confessiones*). It seems he was led to devote himself to the study of philosophy by reading a work of Cicero (*Hortensius*, now lost). He also resorted to the Bible (at that time available in a poor Latin translation, since St. Jeromes's translation would arrive later) and was fascinated by the Manichaean heresy which, with its dualistic conception of the good and the evil, of which for nine years he was one of the more fretful followers.

He began his teaching of rhetoric at first in Carthage, then in Rome and eventually in Milan. There, after having heard the sermons of St. Ambrose, he was definitively converted to Christianity. He gave up teaching and decided to go back to Africa and devote himself to a life of devoutness and study. His mother's death in Ostia, before his departure, kept him in Rome for one more year. By 388 he was already in Tagaste, where he lived with his friends for three years in a meditative idleness. In 392 he entered the priesthood and, after having been ordained auxiliary bishop by Valerius, at Valerius' death (in 396) he was elected bishop of Hippo, where he remained until his death in 430. These sketchy biographical notes are only given in order to fix the dates in the historical excursus we are making. Obviously, one cannot expect to summarize the life of a philosopher as outstanding as St. Augustine in a few lines.

For our purposes, we shall deal with one of his minor works, which dates back to the last years of his life: *De Genesi ad litteram*. As is evident from the title, this work deals with a problem we have already faced: the conflict between the Bible and the cosmography of Hellenistic science. Augustine was already presented the problem in Book XII of the *Confessiones*, where he discourses on the meaning of the two firsts verses, which affirm that God created the heavens and the Earth and that the Earth was void and empty, and then continues until "the supercelestial waters" (*Confessiones*, XII, 21–22). But in *De Genesi ad litteram*, the discussion is more accurate and thorough (obviously, from Augustine's point of view), therefore, we shall tackle some passages of this work in order to clarify Augustine's opinion as to the problem. Let us start from the beginning, where Augustine opposes those who hold that the waters cannot stay above the heaven:

> For now it is our business to seek in the account of Holy Scripture how God made the universe, not what He might produce in nature or from nature by His miraculous power. If God ever wished oil to remain under water, it would do so. But we should not thereby be ignorant of the nature of oil: we should still know that it is so constituted as to tend towards its proper place and, even when poured under water, to make its way up and settle on the surface. Now we are seeking to know whether the Creator, who has *ordered all things in measure, and number, and weight,* has assigned to the mass of waters not just one proper place around the earth, but another also above the heavens, a region which has been spread around and established beyond the limits of the air.
>
> Those who deny this theory base their argument on the weights of the elements. Surely, they say, there is no solid heaven laid out above like a pavement to serve as a support for the mass of water. Such a solid body, they argue, cannot exist except on the earth, and whatever is so constituted is earth, not heaven. They go on to show that the elements are distinguished not by their locations only but also by their qualities, and that each is assigned its place in keeping with its particular qualities.[36]

Further on, he goes into explanations and hazardous justifications which have little to do with the real physics of planetary motion:

> Certain writers, even among those of our faith, attempt to refute those who say that the relative weights of the elements make it impossible for water to exist above the starry heaven. They base their arguments on the properties and motions of the stars. They say that the star called Saturn is the coldest star, and that it takes thirty years to complete its orbit in the heavens because it is higher up and therefore travels over a wider course. The sun completes a similar orbit in a year, and the moon in a month, requiring a briefer time, they explain, because these bodies are lower in the heavens; and thus the extent of time is in proportion to the extent of space.
>
> These writers are then asked why Saturn is cold. Its temperature should be higher in proportion to the rapid movement it has by reason of its height in the heavens. For surely when a round mass is rotated, the parts near the centre move more slowly, and those near the edge more rapidly, so that the greater and lesser distances may be covered simultaneously in the same circular motion. Now, the greater the speed of an object, the greater is

[36]St. Augustine: *De Genesi ad litteram*, II, 1: *Nunc enim quemadmodum Deus instituerit naturas rerum, secundum Scripturas eius nos convenit quaerere; non quid in eis vel ex eis ad miraculum potentiae suae velit operari. Neque enim si vellet Deus sub aqua oleum aliquando manere, non fieret; non ex eo tamen olei natura nobis esset incognita, quod ita facta sit, ut appetendo suum locum, etiam si subterfusa fuerit, perrumpat aquas, eisque se superpositam collocet. Nunc ergo quaerimus utrum conditor rerum, qui omnia in mensura et numero et pondere disposuit, non unum locum proprium ponderi aquarum circa terram tribuerit, sed et super coelum quod ultra limitem aeris circumfusum atque solidatum est.*

Quod qui negant esse credendum, de ponderibus elementorum argumentantur, negantes ullo modo ita desuper quasi quodam pavimento solidatum esse coelum, ut possit aquarum pondera sustinere; quod talis soliditas nisi terris esse non possit, et quidquid tale est, non coelum sed terra sit. Non enim tantum locis, sed etiam qualitatibus elementa distingui, ut pro qualitatibus propriis etiam loca propria sortirentur (Eng. trans. John Hammond Taylor).

heat. Accordingly, this star ought to be hot rather than cold. It is true, indeed, that by its own motion, moving over a vast space, it takes thirty years to complete its orbit; yet by the motion of the heavens it is rotated rapidly in the opposite direction and must daily travel this course (and thus, they say, each revolution of the heavens accounts for a single day); and, therefore, it ought to generate greater heat by reason of its greater velocity. The conclusion is, then, that it is cooled by the waters that are near it above the heavens, although the existence of the waters is denied by those who propose the explanation of the motion of the heavens and the stars that I have briefly outlined.

With this reasoning some of our scholars attack the position of those who refuse to believe that there are waters above the heavens while maintaining that the star whose path is in the height of the heavens is cold. Thus the would compel the disbeliever to admit that water is there not in a vaporous state but in the form of ice. But whatever the nature of that water and whatever the manner of its being there, we must not doubt that it does exist in that place. The authority of Scripture in this matter is greater than all human ingenuity.[37]

He therefore concludes that, no matter what explanation is given, the authority of Scripture is in any case superior to the capacity of whatever human mind. The final topic is the spherical shape of the heavens, which is in any case not so important, as this knowledge is not essential for the salvation of the soul and the beatitude of the faithful:

It is also frequently asked what our belief must be about the form and shape of heaven according to Sacred Scripture.

Many scholars engage in lengthy discussions on these matters, but the sacred writers with their deeper wisdom have omitted them. Such subjects are of no profit for those who seek beatitude, and, what is worse, they take up very precious time that ought to be given to what is spiritually beneficial. What concern is it of mine whether heaven is like a sphere and the earth is enclosed by it and suspended in the middle of the universe, or whether heaven like a disk above the earth covers it over on one side?

[37]St. Augustine: *De Genesi ad litteram*, II, 5: *Quidam etiam nostri, istos negantes propter pondera elementorum aquas esse posse super coelum sidereum, de ipsorum siderum qualitatibus et meatibus convincere moliuntur. Iidem namque asserunt stellam quam Saturni appellant, esse frigidissimam, eamque per annos triginta signiferum peragere circulum, eo quod superiore ac per hoc ampliore ambitu graditur. Nam sol eumdem circulum per annum complet, et luna per mensem; tanto, ut dicunt, brevius, quanto inferius, ut spatio loci spatium temporis congruat. Quaeritur itaque ab eis, unde illa stella sit frigida, quae tanto ardentior esse deberet, quanto sublimiore coelo rapitur. Nam procul dubio cum rotunda moles circulari motu agitur, interiora eius tardius eunt, exteriora celerius, ut maiora spatia cum brevioribus ad eosdem gyros pariter occurrant: quae autem celerius, utique ferventius. Proinde memorata stella magis debuit calida esse quam frigida: quamvis enim suo motu, quoniam grande spatium est, triginta annis totum ambitum permeet, tamen coeli motu in contrarium rotata velocius, quod quotidie necesse est patiatur (sic, ut dicunt, coeli singulae conversiones, dies singulos explicant), calorem maiorem debuit coelo concitatiore concipere. Nimirum ergo eam frigidam facit aquarum super coelum constitutarum illa vicinitas, quam nolunt credere, qui haec, quae breviter dixi, de motu coeli et siderum disputant. His quidam nostri coniecturis agunt adversus eos qui nolunt aquas super coelum credere, et volunt eam stellam esse frigidam, quae iuxta summum coelum circuit; ut ex hoc cogantur aquarum naturam, non iam illic vaporali tenuitate, sed glaciali soliditate pendere. Quoquo modo autem et qualeslibet aquae ibi sint, esse eas ibi minime dubitemus: maior est quippe Scripturae huius auctoritas, quam omnis humani ingenii capacitas* (Eng. trans. John Hammond Taylor).

But the credibility of Scripture is at stake, and as I have indicated more than once, there is danger that a man uninstructed in divine revelation, discovering something in Scripture or hearing from it something that seems to be at variance with the knowledge he has acquired, may resolutely withhold his assent in other matters where Scripture presents useful admonitions, narratives, or declarations. Hence, I must say briefly that in the matter of the shape of heaven the sacred writers knew the truth, but the Spirit of God, who spoke through them, did not wish to teach men these facts that would be of no avail for their salvation.

But someone may ask: "Is not Scripture opposed to those who hold that heaven is spherical, when it says, *who stretches out heaven like a skin*?" Let it be opposed indeed if their statement is false. The truth is rather in what God reveals than in what groping men surmise. But if they are able to establish their doctrine with proofs that cannot be denied, we must show that this statement of Scripture about the skin is not opposed to the truth of their conclusions. If it were, it would be opposed also to Sacred Scripture itself in another passage where it says that heaven is suspended like a vault. For what can be so different and contradictory as a skin stretched out flat and the curved shape of a vault? But if it is necessary, as it surely is, to interpret these two passages so that they are shown not to be contradictory but to be reconcilable, it is also necessary that both of these passages should not contradict the theories that may be supported by true evidence, by which heaven is said to be curved on all sides in the shape of a sphere, provided only that this is proved.[38]

2.3.4 *The Last Latin Encyclopedists of the Roman Empire*

We have already pointed out that in the Roman world scientific knowledge was not always widespread, at a given time, in a uniform way, which allows one to suppose

[38]St. Augustine: *De Genesi ad litteram*, II, 9: *Quaeri etiam solet quae forma et figura coeli esse credenda sit secundum Scripturas nostras. Multi enim multum disputant de iis rebus, quas maiore prudentia nostri auctores omiserunt, ad beatam vitam non profuturas discentibus; et occupantes, quod peius est, multum pretiosa, et rebus salubribus impendenda temporum spatia. Quid enim ad me pertinet, utrum coelum sicut sphaera undique concludat terram in media mundi mole libratam, an eam ex una parte desuper velut discus operiat? Sed quia de fide agitur.*

*Scripturarum, propter illam causam, quam non semel commemoravi, ne quisquam eloquia divina non intellegens, cum de his rebus tale aliquid vel invenerit in Libris nostris, vel ex illis audierit, quod perceptis a se rationibus adversari videatur, nullo modo eis caetera utilia monentibus, vel narrantibus, vel pronuntiantibus credat; breviter dicendum est de figura coeli hoc scisse auctores nostros quod veritas habet; sed Spiritum Dei, qui per ipsos loquebatur, noluisse ista docere homines nulli saluti profutura. Sed, ait aliquis, quomodo non est contrarium iis qui figuram sphaerae coelo tribuunt, quod scriptum est in Litteris nostris: Qui extendit coelum sicut pellem? Sit sane contrarium, si falsum est quod illi dicunt: hoc enim verum est quod divina dicit auctoritas, potius quam illud quod humana infirmitas conicit. Sed si forte illud talibus illi documentis probare potuerint, ut dubitari inde non debeat; demonstrandum est hoc quod apud nos de pelle dictum est, veris illis rationibus non esse contrarium: alioquin contrarium erit etiam ipsis in alio loco Scripturis nostris, ubi coelum dicitur velut camera esse suspensum. Quid enim tam diversum et sibimet adversum, quam plana pellis extensio, et camerae curva convexio? Quod si oportet, sicuti oportet, haec duo sic intellegere, ut concordare utrumque, nec sibimet repugnare inveniatur; ita oportet etiam utrumlibet horum illis non adversari disputationibus, si eas forte veras certa ratio declaraverit, quibus docetur coelum sphaerae figura undique esse convexum, si tamen probatu*r (Eng. trans. John Hammond Taylor).

improbable difficulties of communication. As a matter of fact, one cannot speak, at least with regard to the scientific field, of a Roman culture tout court. There are different scientific cultures which coexisted, just as there were different philosophical cultures, whereas the sphericity of the Earth should have been considered an established verity for many years by then.

We have already seen, with the Apologists and the Fathers of the Church, that faithfulness to Scriptures obliged one to deny physical reality. Now instead we shall see the case of three encyclopedists of the final centuries of the Roman Empire who did not dispute the sphericity of the Earth; what is more, one of whom was most probably Christian.

Let us begin, following chronological order, with Calcidius. Nothing is known about this writer except for his work *In Platonis Timaeum Commentarius* (*Commentary on Plato's Timaeus*). No quotations of this work are found in the writings of the authors of late Antiquity, but it began to be well known starting in the twelfth century. At present scholars hold that Calcidius was of the Christian religion, although this was formerly questioned. The work is dedicated to a certain Osius, who in the past was identified as the bishop of Cordoba, who participated in the council of Nicaea in 325, but now this identification is questioned. In any case, scholars agree that the work of Calcidius is typical of the Western fourth century: the philosophy he sets forth in his commentary on *Timaeus* is that called Middle Platonic (slightly delayed with respect to Greek culture of the same period).

The work consists of two parts, the first of which is the translation into Latin of about one half of Plato's *Timaeus* (XVII–LIII).[39] The *Commentary* is prefaced by a sort of letter to the reader in which the author notes the troubles which readers can encounter in interpreting a work generally considered "hard", and explains the way he has followed to make it more understandable to everybody. According to scholars, Calcidius recast into his *Commentary* extensive excerpts of preceding (Greek) works, some of which are now lost (possible authors: Adrastus of Afrodisia, Theon of Smyrna, Porphyrius). He quotes several authors, among whome there is only one Christian: Origen of Alexandria. As to what interests us, in the section *De stellis ratis et errantibus* ("On the fixed and wandering stars") he asserts: "Plato says that the world is round and spherical in shape, that the Earth is spherical and situated at the center of the world, that in its position the Earth is the equivalent of the center…".[40]

Actually, Plato does not assert this so explicitly anywhere: this is an "updating" due to the Platonic scholars of the following centuries. Recalling the problems raised by the verses of Genesis, of which we have spoken regarding Ambrose and Augustine, the reference to Origen that Calcidius makes when dealing with the

[39]We refer the reader to the work Claudio Moreschini: *Calcidio. Commentario al Timeo di Platone* (Testo latino a fronte (2nd ed., Bompiani, 2012) (Italian and Latin facing pages); this is the first translation into a modern language). For the Eng. trans. see Calcidius: *On Plato's Timaeus*, John Magee, trans and ed. (Harvard University Press, 2016).

[40]Calcidius: *Commentarius* LIX: *Ait Plato mundi formam rotundam esse et globosam, terram item globosam in medietate mundi sitam* ... (ed. Moreschini p. 222; Eng. trans. John Magee).

same problem appears somehow astonishing. After having quoted two versions of the first two verses of Genesis, he says:

> But Origen claims to have been persuaded by the Hebrews that the translation deviates slightly from the true sense, and on this grounds that the original had, "but the earth was in a certain state of dumb admiration.[41]

Whilst it appears to have been verified that this version was really common in Rabbinic circles, the fact in any case stands that Calcidius is not concerned with the problem of supercelestial waters. Most probably his references to Hellenistic culture prompted him to address different problems, more philosophic and less physical.

The next encyclopedist we shall deal with is Macrobius (Macrobius Ambrosius Teodosius). As in the case of Calcidius, his biographical notes are very uncertain, but at least a little bit more is known about him. He was not Roman, as he himself informs us in his work *Saturnalia* ("*nos sub alio ortos caelo*"), but he held important positions in the imperial administration at the time of the sons of Theodosius, Honorious and Arcadius (395–423 AD). This circumstance leads one think that he was Christian, since in those times the Christian religion had practically become the state religion, but not all scholars agree on this. Besides the *Saturnalia* (a work which is named after the homonymous Latin feasts and presents a series of conversations that are imagined to have been held at the home of the best-known exponents of the Roman aristocracy), the other important work of Macrobius is *In Somnium Scipionis* (*Commentary of the Dream of Scipio*).[42] This work, which like the *Saturnalia* is dedicated to his son Eustace, is a commentary on that part of the last book of Cicero's *Republic* which we have already dealt with in Sect. 2.1. We recall that Macrobius' commentary has reached us complete and in this way has ensured the preservation of that part of Cicero's *Republic* which contained the Dream of Scipio. As Calcidius behaved earlier towards the *Timaeus*, so does Macrobius behave towards the Dream of Scipio: Macrobius' work is only a pretext for constructing a compendium of Neoplatonic philosophy also containing an extensive treatment of cosmography and geography (eight chapters in the first book and nine in the second). Of course, these are the sections that interest us here.

In passing, we remind that Macrobius' commentary has reached us complete and in this way has ensured the preservation of that part of Cicero's *Republic* which contained the Dream of Scipio. As already Calcidius behaved towards the *Timaeus*, that way does Macrobius towards the Dream of Scipio: Macrobius' work is only a pretext for constructing a compendium of Neoplatonic philosophy also containing an extensive treatment of cosmography and geography (eight chapters in the first book and nine in the second).

---

[41]Calcidius: *Commentarius* CCLXXVI: *Sed Origenes asseverat ita sibi ab Hebraeis esse persuasum quod in aliquantum sit a vera proprietate derivata interpretatio; fuisse enim in exemplari: "Terra autem stupida quadam erat admiratione* (ed. Moreschini p. 564; Eng. trans. John Magee).

[42]See William Harris Stahl (Eng. trans. and ed.): *Commentary on the Dream of Scipio by Macrobius*, 2nd ed. (Columbia University Press, 1990); Moreno Neri (trans. and ed.), *Macrobio. Commento al Sogno di Scipione*, 2nd ed. (Bompiani, 2014) (Latin and Italian facing pages).

In the Middle Ages there were even manuscripts which only contained those chapters by way of a booklet on cosmography. As usual, according to scholars, the work of Macrobius is the result of a composition of different parts copied from different authors (principally from Posidonius and Porphyrius, as it seems). Verifying this is not easy, since one has to do with works now lost, and in that case the check is feasible only indirectly. The dependence on different authors also results in some problems. For instance, regarding the ordering of the planets, he has difficulty in reconciling the opinion of Plato with that of Posidonius. Also, when talking about authors of the previous generations, he sometimes changes his opinion. For instance, the measurement of the terrestrial circumference by Eratosthenes is mentioned—without mentioning the author by name—as obvious and accurate, whereas elsewhere he expresses a negative opinion on Eratosthenes himself:

> By most obvious and accurate methods of measurement we know that the earth's circumference, including the habitable and uninhabitable areas, is 252,000 stades. This being the circumference, the diameter will of course be 80,000 stades, or slightly more, according to the method explained above of tripling the diameter with the addition of a seventh part to get the circumference.[43]

In any case, on the size and the shape of the Earth he is rigorous and reports the result of Hellenistic science directly.

In the second book, we also find an explanation of Crates' terrestrial model, which we discussed in Sect. 1.8:

> [Cicero] in speaking of the inhabited areas as spots and of the earth's inhabitants as cut off, some obliquely, some transversely, and some in a diametrically opposite region, he presented a vivid picture of the spherical nature of the earth. ...
>
> As regards the five belts, I beg you not to think that the two founders of Roman eloquence, Virgil and Cicero, disagree in their views because the latter says that the belts encircle the earth and the former that the belts, which he calls by their Greek name zones, "hold the sky"; a later discussion will prove that they are both correct and still not contradictory. ...
>
> Between the extremities and the middle zone lie two belts which are greater than those at the poles and smaller than the one in the middle, tempered by the extremes of the adjoining belts; in these alone has nature permitted the human race to exist.[44]

---

[43]Macrobius: *Commentary*, I, 20, 20: *Evidentissimis et indubitabilibus dimensionibus constitit uniuersae terrae ambitum, quae ubicumque uel incolitur a quibuscumque uel inhabitabilis iacet, habere stadiorum milia ducenta quinquaginta duo. Cum ergo tantum ambitus teneat, sine dubio octoginta milia stadiorum vel non multo amplius diametros habet, secundum triplicationem cum septimae partis adiectione, quam superius de diametro et circulo regulariter diximus* (Eng. trans. William Harris Stahl).

[44]Macrobius: *Commentary*, II, 5, 5–12: *Et enim maculas habitationum ac de ipsis habitatonbus alios interruptos aduersosque, obliquos etiam et transuersos alios nominando, terrenae sphaerae globositatem tantum non coloribus pinxit. ...*

De quinque autem cingulis ne quaeso aestìmes duorum Romanae facundiae parentum, Maronis et Tullii, dissentire doctrinam, cum toc ipsis cingulis terram redimitam dicat, ille isdem quas Graeco nomine zonas uocat adserat caelum tenen. Vtrumque enim incorruptam ueramque nec alteri contrariarli retulisse rationem procedente disputatìone constabit. ...

Finally as to the question of the antipodes, he writes:

> I can assure you that the uninformed among them think the same thing about us and believe that it is impossible for us to be where we are; they, too, feel that anyone who tried to stand in the region beneath them would fall. But just as there has never been anyone among us who was afraid he might fall into the sky, so no one in their quarter is going to fall upwards since we saw from a previous discussion that all weights are borne by their own inclination towards the earth.[45]

The above passage can be considered both an outline of the passage of Seneca we have already seen (Sect. 2.1), and (more probably) a masked criticism of Lactantius and Augustine.

The last work we want to speak about is *De Nuptiis Philologiae et Mercurii* (*The Marriage of Philology and Mercury*).[46] This work is a bizarre compilation due to a lawyer who was a native Latin speaker resident in Carthage who (according to scholars) flourished in the period 410–439: Martianus Capella. The work consists of nine books (in prose, but with metric inserts) and follows the ancient model of the Menippean satire, drawing inspiration from a famous work (now lost) of Varro: *Disciplinarum libri*. It is an allegorical novel which narrates the marriage of the god Mercury with a very erudite young girl, Philology, with the purpose of presenting the seven liberal arts of the trivium (grammar, dialectic, rhetoric) and the quadrivium (geometry, arithmetic, astronomy, music). The first two books are dedicated to the preparation for the wedding and each of the other seven is dedicated to one of the liberal arts (which, in the ceremony, play the role of the bridesmaids). The work itself is worthless, both from the literary and scientific points of view, but it has turned out to be very important from the documentary point of view since it presents a digest of classical learning as a whole. For this reason, it was a great success in the Middle Ages, and was used as an authoritative source for teaching the liberal arts even in the Cathedral schools, notwithstanding the fact that its author was a pagan. The subject we are concerned with is treated in Book VI, devoted to geometry.

Regarding the sphericity of the Earth, Martianus resorts to the narration of Pytheas of Massalia (whom we have already met in Sect. 1.6) about the island of Thule and the observation which enabled him to assert that the Earth is of spherical shape (*globosam rotunditatis flexibus habendam esse tellurem*). He then continues:

Inter extremos uero et medium, duo maiores ùltimls, medio minores, ex utriusque uidnitatis intemperie temperantur, in hisque tantum uitales auras natura dedit incolis carpere… (Eng. trans. William Harris Stahl).

[45]Macrobius: *Commentary*, II, 5, 26: *Adfirmauerim quoque et apud illos minus rerum peritos hoc aestimare de nobis, nec credere posse nos in quo sumus loco degere, sed opinari, si quis sub pedibus eorum temptaret stare, casurum. Numquam tamen apud nos quisquam timuit ne caderet in caelum, ergo nec apud illos quisquam in supcriora casurus est, sicut «omnia nutu suo pondera" in terram ferri superius relata docuerunt* (Eng. trans. William Harris Stahl).

[46]See William Harris Stahl and E.L. Burge: *Martianus Cappella and the Seven Liberal Arts. Vol. II: The Marriage of Philology and Mercury* (Columbia University Press, 1992).

[596] We must now reveal the size of the earth and what position it holds in the universe. Eratosthenes, a most learned man, used a gnomon in calculating the earth's circumference as 252,000 stadia.

[597] There are bronze hemispherical bowls called scaphia, which mark the passage of the hours by means of a tall, upright stylus, located at the center of the bottom of the bowl. This stylus is called a gnomon. The length of its shadow, measured at the equinox by a determination of its distance from the center, when multiplied twenty-four times, gives the measure of a double circle.

[598] Eratosthenes, upon being informed by official surveyors in the employ of King Ptolemy as to the number of stades between Syene and Meroë, noted what portion of the earth's surface that distance represented; and multiplying according to the proportionate amount, he straightaway determined how many thousands of stadia there were in the earth's circumference.[47]

As unanimously recognized, the explanation given by Capella of the measure of Eratosthenes is quite messy (that of Cleomedes is more clear cut, despite the simplifications), but it has the merit of having handed the correct number down to medieval scholars. Both medieval scholars and Copernicus himself learned from Capella the theory called after Heraclides, that is, that Mercury and Venus rotate around the Sun, whereas the other planets (Sun included) rotate around the Earth.

***

In these first two chapters, we have tried to follow the customary periodization and therefore we have considered as the end of the ancient world, in the West, the fall of the Roman Empire. Undoubtedly, the fall of the last emperor was preceded for some time by the disaggregation both of the state structure and any cultural organisation. The intellectuals of the last centuries of the empire (for instance the three encyclopedists we have mentioned) give the impression of persons who acted as isolated individuals rather than elements of an intellectual milieu. We also have the impression that intellectual life was more vivacious in the North African centres than in Italy. Nevertheless, starting from the age of Constantine, an organisation alternative to that of the state began to form: the Christian Church. Since, due to the limited object of our research, we are particularly interested in inquiring as to the existence in a given age, of a widely-held opinion on the subject "Earth", in what follows we shall pay attention to what the Christian Church asserts on the subject. In fact, in the next centuries the Church will be the only organizer of cultural events and opinion maker.

[47] *The Marriage of Philology and Mercury* VI, 596–598: 596 *Sequitur ut quem mundi locum quamve graditatem sortita sit approbemus. circulus quidem terrae ducentis quinquaginta duobus milibus stadiorum, ut ab Eratosthene doctissimo gnomonica supputatione discussum.* 597 *quippe scaphia dicuntur rotunda ex aere vasa, quae horarum ductus stili in medio fundo siti proceritate discriminant, qui stilus gnomon appellatur, cuius umbrae prolixitas aequinoctio centri sui aestimatione dimensa vicies quater complicata circuli duplicis modum ... reddidit.* 598 *Eratosthenes vero, ab Syene ad Meroen per mensores regios Ptolomaei certus de stadiorum numero redditus, quotaque portio telluris esset advertens, multiplicansque pro partium ratione, circulum mensuramque terrae incunctanter, quot milibus stadiorum ambiretur, absolvit* (Eng. trans. William Harris Stahl).

Suggested Readings

Carrier, R. (2016). *Science education in the early roman empire*. Durham NC: Pitchstone Publishing.

Carrier, R. (2017). *The scientist in the early roman empire*. Durham NC: Pitchstone Publishing.

Heater, P. (2007). *The fall of the roman empire: A new history of rome and the barbarians*. Oxford University Press.

Kulicowski, M. (2016). *The triumph of empire: The roman world from adrian to constantine*. Cambridge MA: Harvard University Press.

Vitruvius. (1960). *The ten books on architecture* (M. H. Morgan, Trans.). New York: Dover.

Chapter 3
The Middle Ages

We shall again follow in this chapter the usual periodization of the historians, according to which one calls Middle Ages the period of about ten centuries which goes from the fall of the Western Roman Empire to the end of the fifteenth century. Obviously, on one hand this periodization is the child of a Eurocentric point of view, but on the other hand the subject we are dealing with is also inserted in the context of the cultural development (or regression) of the Western world (which, at least until the sixteenth century, corresponded to Europe).

We are also giving way to the temptation of considering the opinion on the sphericity of the Earth as a sort of thermometer for measuring the unscientific fever in Christian Europe. This approach may seem excessive, but it should be noted that the legitimation of the Christian religion, and its subsequent recognition as the established religion, made sure that natural philosophy was given short shrift: man was more important than nature.

The foremost problem was the safety of the soul, since all that concerned the external world was written in Genesis. In this connection, we remark that St. Augustine, a cultural intellectual and profoundly knowledgeable about classical culture, had always maintained an ambivalent behavior toward the latter, and in his work, written four years before his death (see *Retractionum libri duo*, I, 4, 4), ended by expressing his regret for having extolled, in the past, the study of the seven liberal arts and by concluding that the natural sciences are not helpful to the Christian. St. Augustine is the philosopher who exerted the greatest influence on Christian theologians in the Early Middle Ages.

In any case, leaving aside the attitude of the Christian Church, in the western part of the Roman Empire the knowledge of the Greek language decreased more and more, until it remained the inheritance of a very few intellectuals in the sixth century (e.g., Boethius). Since the language of science had been Greek (only the encyclopaedias which popularized the breakthroughs of the Greek science were written in Latin), already by the Early Middle Ages nobody could gain access to the Greek originals any more: scholars were obliged to make the best of the Latin popularizations.

D. Boccaletti, *The Shape and Size of the Earth*,
https://doi.org/10.1007/978-3-319-90593-8_3

As we shall see, Boethius, who worked out a plan of translation of the most important works of Greek philosophy, did not have enough time to translate the fundamental texts of the natural philosophy. It was necessary to wait until the twelfth and thirteenth centuries for Latin translations of the works that the Arabs had translated from the Greek in the ninth and tenth centuries.

3.1 Boethius: The End of the Latin World and the Beginning of the Middle Ages

During the second half of the fifth century, the fall of the Western Roman Empire was both preceded and followed by its general fragmentation into the so-called Roman Barbarian Kingdoms. Italy, at the end of the fifth century and in the first three decades of the sixth, experienced a period of relative quiet, despite the inner tensions in the relations between the Romans and the Goths and the outer tensions in the relations with the Eastern Empire: this is the period of the Gothic Kingdom of Theodoric the Great.

It is in this period that the public and scholarly life of Boethius takes place. Anicius Manlius Torquatus Severinus Boethius was born in Rome in about 480 into a senatorial family and was consul and senator under the Goth-Roman Kingdom. His career as a high official in the kingdom of Theodoric was favoured by the admiration aroused by his studies, appreciated in a series of letters by Cassiodorus, secretary to Theodoric. As a scholar of Greek philosophy, Boethius worked out a programme of translation of the main works (above all of Aristotle and Plato), also maintaining a personal position of synthesis between Platonism and Aristotelianism. The political vicissitudes prevented him from accomplishing his programme which, as regards Aristotle's works, ended with the works of logic.

At the acme of his political career, in 523, Boethius was involved in a conspiracy and accused of treason for having defended the patrician Albinus, accused of having conspired in support of the Eastern emperor. He was imprisoned in Pavia under the accusation of treason, sacrilege and magic. In prison, Boethius, in waiting for the outcome of the trial, wrote his fundamental work: *De consolatione philosophiae*. At the end of the trial he was sentenced to death and killed by order of Theodoric in 525. His work as a philosopher was cut short by the death sentence, but even so his influence in the Middle Ages was notable with regard to the studies of both logic and the arts of the quadrivium (among these, rediscovered in the Carolingian age, the works *De institutione musica* and *De institutione aritmetica* survived).

He wrote several other works (both original works and commentaries), but we are particularly interested in *De consolatione philosophiae*. It is in this work (in prose with inserts in verse) that Boethius succeeds in translating the mystery of the Christian creation into a Platonic conception following the model of the *Timaeus*. The work presents itself as a dialogue, in the course of which Philosophy, which had suddenly appeared to the writer in a vision, shows him that the misfortunes that have befallen him, as those of all the others, are inserted in a reality governed for the better by Providence.

In the work there is no explicit precise reference to the shape and the size of the Earth, but the Platonic cosmography turns out to be the model followed and the God of Genesis takes shape in this work in specifically philosophical terms (that is, in terms of Platonic philosophy). The Boethian synthesis between Platonism, Aristotelianism (the still mover) and the Christian doctrine appears in all evidence in the following poetic text:

> Thou who dost rule the universe with everlasting law, founder of earth and heaven alike, who hast bidden time stand forth from out Eternity, for ever firm Thyself, yet giving movement unto all. No causes were without Thee which could thence impel Thee to create this mass of changing matter, but within Thyself exists the very idea of perfect good, which grudges naught, for of what can it have envy? Thou makest all things follow that high pattern. In perfect beauty Thou movest in Thy mind a world of beauty, making all in a like image, and bidding the perfect whole to complete its perfect functions. All the first principles of nature Thou dost bind together by perfect orders as of numbers, so that they may be balanced each with its opposite: cold with heat, and dry with moist together; thus fire may not fly upward too swiftly because too purely, nor may the weight of the solid earth drag it down and overwhelm it. Thou dost make the soul as a third between mind and material bodies: to these the soul gives life and movement, for Thou dost spread it abroad among the members of the universe, now working in accord. Thus is the soul divided as it takes its course, making two circles, as though a binding thread around the world. Thereafter it returns unto itself and passes around the lower earthly mind; and in like manner it gives motion to the heavens to turn their course. Thou it is who dost carry forward with like inspiration these souls and lower lives. Thou dost fill these weak vessels with lofty souls, and send them abroad throughout the heavens and earth, and by Thy kindly law dost turn them again to Thyself and bring them to seek, as fire doth, to rise to Thee again. Grant then, O Father, that this mind of ours may rise to Thy throne of majesty; grant us to reach that fount of good. Grant that we may so find light that we may set on Thee unblinded eyes; cast Thou therefrom the heavy clouds of this material world. Shine forth upon us in Thine own true glory. Thou art the bright and peaceful rest of all Thy children that worship Thee. To see Thee clearly is the limit of our aim. Thou art our beginning, our progress, our guide, our way, our end.[1]

---

[1]Boethius, *Consolatio*, III, IX: *O qui perpetua mundum ratione gubernas,/ terrarum caelique sator, qui tempus ab aevo/ ire iubes stabilisque manens das cuncta moveri,/ quem non externae pepulerunt fingere causae/materiae fluitantis opus, verum insita summi/forma boni livore carens, tu cuncta superno/ducis ab exemplo, pulchrum pulcherrimus ipse/mundum mente gerens similique in imagine formans/perfectasque iubens perfectum absolvere partes./ Tu numeris elementa ligas, ut frigora flammis,/ arida conveniant liquidis, ne purior ignis/ evolet aut mersas deducant pondera terras./ Tu triplicis mediani naturae cuncta moventem/conectens animam per consona membra resolvis;/ quae cum secta duos motum glomeravit in orbes,/ in semet reditura meat mentemque profundam/circuit et simili convertit imagine caelum./ Tu causis animas paribus vitasque minores/ provehis et levibus sublimes curribus aptans/in caelum terramque seris, quas lege benigna/ad te conversas reduci facis igne reverti./ Da, pater, augustam menti conscendere sedem,/ da fontem lustrare boni, da luce reperta/in te conspicuos animi defigere visus./ Dissice terrenae nebulas et pondera molis/atque tuo splendore mica; tu namque serenum,/ tu requies tranquilla piis, te cernere finis,/ principium, vector, dux, semita, terminus idem* (Eng. trans. H. R. James). This poem is undoubtedly the most important of the Consolatio and constitutes the heart of the whole treatise.

As is known, Boethius has been described as the last of the Romans and the first of the Scholastics. In fact, even if detached from the golden age of Roman classical literature, he closes the Roman cultural tradition, as in another field Ptolemy (he too delayed) closed that of Hellenistic science.

3.2 The Christians Encyclopedists of the Early Middle Ages

At the beginning of the seventh century, which opens the period we shall now discuss, the territory of the Roman Empire was almost steadily subdivided into many so-called barbarian kingdoms, which also became territories of evangelization. In the course of time, in fact, several preachers of the Catholic religion scattered throughout the rest of Western Europe from Rome and, in their turn, had set up evangelization movements on a local basis which, in time, assumed particular features.

In these contexts several writers flourished (in general, they were bishops or important members of the local churches) who, in addition to works of theology of a strictly religious character, also wrote works which took the place of the compilations which had circulated in the Roman world in the past. Thereby, a new generation of encyclopedists came into being: the Christian encyclopedists. As we shall see, their works (all written in Latin) were gradually diffused throughout the entire Christian world (again we referring entirely to the West), and we can say that their basic philosophical approach was reminiscent of Neoplatonism.

3.2.1 Isidore of Seville

For a very long time Spain, which in the past had been one of the more radically Latinized regions (it had produced both emperors and great intellectuals), was one of those where the evangelization had achieved success (in the period we are dealing with). One of the several bishoprics inside the Visigothic Kingdom of Spain was Seville, and there Isidore was born in 560. Born into a noble family, Isidore appears in the theological-literary arena in 583 by participating in the resumption of the fight against the Arians. From 601 until his death in 636 he was bishop of Seville, succeeding his older brother Leander. In his activity as bishop, he chaired several councils, among them the Fourth Council of Toledo in 633 where (a rare boast for a Spanish bishop) he deplored the forced conversion of Jews to Christianity imposed by King Sisebute (612–621). However, Isidore's activity as a bishop is not his most distinguishing characteristic; rather it is his irrepressible fertility as a writer, which made him the most influential and universally widespread author of the Christian Middle Ages. Nevertheless, he never reached the summit of either literary art or philosophical inquiry, but was essentially a compilator: he took upon himself the task of collecting all that had reached him of ancient culture and handing it down to posterity. Modern opinions on the quality of his work are not favourable. For instance, Gabriele Pepe wrote that "more than an encyclopedist he is an intrepid copyist", while Lynn Thorndike said, "the work's importance consists chiefly in showing how scanty was the knowledge of the early middle ages".[2] Nonetheless, his work remains fundamental since it covers all the fields of interest in his time and the information he gives is the most complete. We add that even Dante Alighieri mentioned Isidore in his *Divina Commedia* ("*l'ardente spiro di Isidoro*", *Paradiso*, X, 131) and, in our own day, in 2002 Pope John Paul II designated him patron saint of the Internet.

Among the huge number of his works, two are of interest for us: *Etymologiae* (sometimes also called *Origines*) and *De natura rerum*. The first work, in 20 books, is a vast collection of notions on the most varied subjects, based on the etymology of the words (conceived following the opinions of that time and then very often quite fanciful). Book XIV is dedicated to the Earth and its parts. *De natura rerum*, dedicated to King Sisebute, is a little work in 48 chapters; it ranges from astronomy to meteorology and geography. The style of these works is sober and clear since, as Isidore says in another work, "A prolix and obscure speech bores: a concise and open one delights".[3]

After this, it is natural to wonder, as Ernest Brehaut does, "Is it possible to ascertain from the writings of Isidore what was the general view of the universe and the attitude toward life held in the sixth and seventh centuries?" His answer is:

> On first thought it seems doubtful. As has been indicated, his works, and especially the Etymologies, form a mosaic of borrowings, whose ultimate origin is to be traced to unnumbered writings in both Greek and Latin, and in both Christian and pagan literatures. We find side by side in Isidore the ideas of Aristotle, Nicomachus, Porphyry, Varro, Cicero, Suetonius, Moses, St. Paul, Origen, and Augustine, to mention only a few; and these ideas, although as a rule they have undergone degeneration, are sometimes in the original words or a close rendering of them. If viewed closely they are a mass of confusion and incoherence.[4]

In any case, one must take into account that Isidore came after centuries of compilations by the Roman encyclopedists, in which the inner coherence was not the greatest value, since the necessary condition for making a good divulgation is to master the matter to be divulgated and this was not the case of the Roman compilators nor of Isidore, at least for what concerns the natural sciences. Obviously, we are interested in ascertaining what Isidore's world view was; as we shall see, there is little or nothing new with reference to Ambrose and Augustine.

---

[2]See G. Pepe: Il Medioevo barbarico in Italia (Il Saggiatore, 1967), p. 468; Lynn Thorndyke, *A History of Magic and Experimental Science* (Columbia University Press, 1923), vol. 1, p. 623.

[3]Isidore, *Quaestiones in vetus testamentum. In Genesin.* Praefatio, 3: *Prolixa enim et occulta taedet oratio, brevis et aperta delectat*" (our Eng. trans.).

[4]Ernest Brehaut: *An Encyclopedist of the Dark Ages: Isidore of Seville* (Columbia University, 1912), p. 48.

Let us begin with an excerpt from Book XIV of *Etymologiae*:

Chapter 1. On the earth

1. The earth is placed in the middle region of the universe, being situated like a centre at an equal interval from all parts of heaven; in the singular number it means the whole circle; in the plural the separate parts; and reason gives different names for it; for it is called *terra* from the upper part where it suffers attrition; *humus* from the lower and humid part, as for example, under the sea; again, *tellus,* because we take its fruits; it is also called *ops* because it brings opulence. It is likewise called *arva,* from ploughing and cultivating.

2. Earth in distinction from water is called dry; since the Scripture says that "God called the dry land, earth". For dryness is the natural property of earth. Its dampness it gets by its relation to water. As to its motion (earthquakes) some say it is wind in its hollow parts, the force of which causes it to move.

3. Others say that a generative water moves in the lands, and causes them to strike together, *sicut vas,* as Lucretius says. Others have it that the earth is sponge-shaped, and its fallen parts lying in ruins cause all the upper parts to shake. The yawning of the earth also is caused either by the motion of the lower water, or by frequent thunderings, or by winds bursting out of the hollow parts of the earth.

Chapter 2. On the circle of lands

1. The circle of lands is so called from its roundness, which is like that of a wheel, whence a small wheel is called *orbiculus.* For the Ocean flowing about on all sides encircles its boundaries. It is divided into three parts; of which the first is called Asia; the second, Europe; the third, Africa.

2. These three parts the ancients did not divide equally; for Asia stretches from the South through the East to the North, and Europe from the North to the West, and thence Africa from the West to the South. Whence plainly the two, Europe and Africa, occupy one-half, and Asia alone the other. But the former were made into two parts because the Great Sea enters from the Ocean between them and cuts them apart. Wherefore if you divide the circle of lands into two parts, East and West, Asia will be in one, and in the other, Europe and Africa.[5]

[5]Isidore of Seville, *Isidori Hispalensis Episcopi Etymologiarum sive Originum Liber XIV,* 1–2: *Terra est in media mundi regione posita, omnibus partibus caeli in modum centri aequali intervallo consistens; quae singulari numero totum orbem significat, plurali vero singulas partes. Cuius nomina diversa dat ratio; nam terra dicta a superiori parte, qua teritur; humus ab inferiori vel humida terra, ut sub mari; tellus autem, quia fructus eius tollimus; haec et Ops dicta, eo quod opem fert frugibus; eadem et arva, ab arando et colendo vocata. Proprie autem terra ad distinctionem aquae arida nuncupatur, sicut Scriptura ait (Genes. 1, 10): "Quod vocaverit Deus terram aridam." Naturalis enim proprietas siccitas est terris; nam ut humida sit, hoc aquarum affinitate sortitur. Cuius motum alii dicunt ventum esse in concavis eius, qui motus eam movet. Sallustius (Hist. 2, fr. 28): "Venti per cava terrae citatu rupti aliquot montes tumulique sedere."* Alii *aquam dicunt genetalem in terris moveri, et eas simul concutere, sicut vas, ut dicit Lucretius (6, 555). Alii* σπογγοειδῆ *terram volunt, cuius plerumque latentes ruinae superposita cuncta concutiunt. Terrae quoque hiatus aut motu aquae inferioris fit, aut crebris tonitruis, aut de concavis terrae erumpentibus ventis. Orbis a rotunditate circuli dictus, quia sicut rota est; unde brevis etiam rotella orbiculus appellatur. Undique enim Oceanus circumfluens eius in circulo ambit fines. Divisus est autem trifarie: e quibus una pars Asia, altera Europa, tertia Africa nuncupatur. Quas tres partes orbis veteres non aequaliter diviserunt. Nam Asia a meridie per orientem usque ad septentrionem pervenit; Europa vero a septentrione usque ad occidentem; atque inde Africa ab occidente usque ad meridiem. Unde evidenter orbem dimidium duae tenent. Europa et Africa, alium vero dimidium sola Asia; sed ideo istae duae partes factae sunt, quia inter utramque ab Oceano mare Magnum*

As appears evident, Isidore's Earth is a disc (*sicut rota est*) coincident with the oecumene, of which he enumerates the nature of the parts. The Earth has a central position in the universe and this is the only "cosmographic" statement. Let us see now what he states in *De natura rerum*. Isidore considers the universe bounded by a revolving sphere which he belives to be made of fire and in which the stars are fixed (but, alternatively, he also talks of seven spheres).

At the end of Chap. XIII, he says: "Then he solidified the heaven of the lower orb not only by uniformity but by a multiplicity of motions, calling it the firmament because it sustains the higher waters."[6] Then then continues in Chap. XIV:

> 1. This thought is from Ambrose: "Wise men of the world say that waters cannot be over the heavens, saying that if the heaven is fire and the nature of water is not able to mix with it. They add to which saying the orb of the sky is round and voluble and warm and in that voluble circle water can never stand still. For it is necessary that they [waters] flow and glide when it [heaven] is turned from a higher to a lower orb, so for this reason they can never stand still there because the axis of the sky turning and revolving pours them forth with swift motion.
>
> 2. But they are being nonsensical and speak confusedly, because Whoever can create something from nothing can set waters in the sky with the nature of solid ice. For when they say the shining orb of the heaven revolves with burning stars, divine providence necessarily looks on, so that between the orb of the heaven water flows and the heat of the burning axis is tempered.[7]

As one can see in the second part of the last excerpt, even calm Isidore is carried away by passion after the fashion of Lactantius; only the question of the Antipodes is missing!

---

ingreditur, quod eas intersecat. Quapropter si in duas partes orientis et occidentis orbem dividas, Asia erit in una, in altera vero Europa et Africa (Eng. trans. Ernest Brehaut, *An Encyclopedist of the Dark Ages*, op. cit., pp. 243–244).

[6]Isidore of Seville, *Hisidori Hispalensis De natura rerum*, XIII, 1: *Dehinc circulum inferioris coeli, non uniformi, sed multiplici motu solidavit, nuncupans eum firmamentum propter sustentationem superiorum aquarum* (Eng. trans. Carolyn Embach).

[7]Isidore of Seville, *Hisidori Hispalensis De natura rerum*, op. cit., XIV, 1–2: *1 Haec est Ambrosii sententia: "Aquas super coelos sapientes mundi hujus aiunt esse non posse, dicentes: igneum esse coelum, non posse concordari eum eo naturam aquarum. Addunt quoque, dicentes rotundum, ac volubilem, atque ardentem esse orbem coeli, et in illo volubili circuitu aquas stare nequaquam posse. Nam necesse est, ut defluant, et labantur, eum de superioribus ad inferiora orbis ille detorquetur, ac per hoc nequaquam eas stare posse aiunt, quod axis coeli concito se motu torquens eas volvendo effunderet". 2 Sed hi tandem insanire desinant, atque confusi agnoscant, quia qui potuit cuncta creare ex nihilo, potuit et illam aquarum naturam glaciali soliditate stabilire in coelo. Nam cum et ipsi dicant volvi orbem stellis ardentibus refulgentem, nonne divina Providentia necessario prospexit, ut inter orbem coeli redundarent aquae, quae illa ferventis axis incendia temperarent?* (Eng. trans. Carolyn Embach).

3.2.2 Bede the Venerable

As we have already mentioned, after the fall of the Roman Empire, Christian evangelization had propagated the religion of Rome a little bit everywhere in the various barbaric kingdoms of central and northern Europe. One noteworthy region from this point of view was Britain. Here, almost a century after Isidore Bede was born (in 672 or 673), precisely in the Kingdom of Northumbria, on Britain's eastern coast. Here he spent all his life as a monk in the two "twin" monsteries of Wearmonth and Yarrow, as he himself says in the autobiographical notes provided in the final part of his most important work, *Historia ecclesiastica gentis Anglorum*, V, 24, 2), entirely devoted to prayer, intense and prolific study, and teaching. According to historians, he was the most important intellectual of his time and wrote works of history, theology and science. Of course, not all historians agree. For instance Lynn Thorndike was not so generous: "Bede perhaps knew more natural science than anyone of his time, but if so, the others must have known practically nothing; his knowledge can in no sense be called extensive".[8] In addition to the *Historia ecclesiastica gentis Anglorum*, Bede also wrote works which can be included in the tradition of the Christian encyclopedists, such as Isidore.

One of the most discussed questions in the Christian field at that time (and for many centuries to come!) was the determination of the date of Easter (the task was to determine a date in the Julian Calendar starting from the story told in the Gospel which was based on references to a lunar calendar). Bede was certainly the most competent author on the subject and tried to explain the thing to his fellows in an accessible way, since he wrote not "as a critic for critics but as a student of sacred literature whose object is the instruction and edification of devote minds", as Claude Jenkins writes.[9] Bede devoted to this subject a first work, *De temporibus*, followed in the maturity by *De temporum ratione*. At almost the same time with the first, he also wrote the little work *De natura rerum*, which is the one of interest for us, for obvious reasons. *De natura rerum* is a concise booklet made up of 51 chapters in which the fundamental notions on the nature and the structure of the universe are presented. Apart from the initial part, it follows Isidore's homonymous work, even while differing from it in certain conclusions. The "scientific" ideas are more often derived from Pliny, where the *Naturalis historia* prevails over Isidore. Bede also follows Pliny in his blunders (for instance: the size of the Moon in comparison with that of the Earth). But most important for us, Bede follows Pliny as regards the shape of the Earth. As Lynn Thorndike, an expert in medieval manuscripts, remarks:

---

[8]Lynn Thorndike: *A History of Magic and Experimental Science*, op. cit., vol. I, p. 634.

[9]C. Jenkins: *Bede as exegete and theologian in Bede, His life, times and writings, Essays in commemoration of the twelfth centenary of his death*, 2nd ed. (Oxford Clarendon Press, 1969), p. 170.

In addition to Bede's own statement of his aim, the frequency with which we find manuscripts of early date of the *De natura rerum* and *De temporibus* suggests they were employed as textbooks in the monastic schools of the early middle ages.[10]

Therefore, it is reasonable to think that these two little works had a notable impact on the formation of ideas regarding the matters we are dealing with in the only environment in which at that time learning was circulating.

Let us focus on *De natura rerum.* While if on one hand the utmost schematization pursued by Bede did not allow the learner monk to broaden his knowledge of a subject, on another hand it allows us to clearly understand Bede's thought regarding certain subjects. Let us begin with the description of the world, which opens with two brief chapters on the creation. The third chapter is:

3. What the World Is

The world is the entire universe, which consists of heaven and earth, rounded out of four elements into the appearance of a complete sphere: out of fire, by which the stars shine; out of air, by which all living things breathe; out of the waters, which barricade the earth by surrounding and penetrating it; and out of earth itself, which is the middle and lowest portion of the world. It hangs suspended, motionless, with the universe whirling around it. But heaven is also called by the word 'mundus', meaning 'elegant', from its perfect and absolute elegance; for it is called 'cosmos' by the Greeks from its adornment.[11]

Here we have, essentially, a Bible-*Timaeus* synthesis, that is, a Platonic cosmology inserted into the biblical story of creation. The cosmological talk continues in the fifth chapter, of which we give an excerpt:

5. The Firmament

Heaven is of a fine and fiery nature, and round and arranged on all sides at equal distances from the centre of the earth. Hence it appears to be vaulted and centred wherever it may be viewed. Those who are knowledgeable about the world have stated that it revolves daily with indescribable swiftness, so that it would destroy itself if it were not restrained by the countervailing course of the planets.[12]

Let us pass on, finally, with the *vexata quaestio* of the celestial waters, dealt with in the eighth chapter:

---

[10]Lynn Thorndike, *A History of Magic and Experimental Science*, op. cit., vol. I, pp. 634–635.

[11]Bede, *De Natura rerum, caput* III: *Quid sit mundus. Mundus est universitas omnis, quae constat ex coelo et terra, quatuor elementis in speciem orbis absoluti globata: igne, quo sidera lucent; aere, quo cuncta viventia spirant; aquis, quae terram cingendo et penetrando communiunt; atque ipsa terra, quae mundi media atque ima, librata volubili circa eam universitate pendet immobilis. Verum mundi nomine etiam eoelum a perfecta absolutaque elegantia vocatur; nam et apud Graecos ab ornatu χόσμος appellatur* (Eng. trans. Bede: On the Nature of Things and On Times, trans. Calvin B. Kendall and Faith Wallis (Liverpool: Liverpool University Press, 2010), p. 75).

[12]Bede, *De Natura rerum, caput* V: *De firmamento. Coelum subtilis igneaeque naturae, rotundumque, et a centro terrae aequis spatiis undiquc collectum. Unde et convexum mediumque, quacunque cernatur, inenarrabili celerilate quotidie circumagi sapientes mundi dixerunt, ita ut rueret, si non planetarum occursu moderaretur…* (Eng. trans. Calvin B. Kendall and Faith Wallis, op. cit., p. 76).

8. The Heavenly Waters

Some people maintain that the waters placed above the firmament, lower indeed than the spiritual heavens but nevertheless superior to every corporeal creation, were reserved for the inundation of the Flood, but others claim more correctly that they were suspended to temper the first of the stars.[13]

It is undoubted according to Bede that the celestial waters exist, but he deals with the subject as briefly as possible by quoting two justifications of their exixtence.

Let us see now that, as we have already anticipated, Bede follows Pliny when dealing with the shape of the Earth:

45. The Position of the Earth

The earth is fixed on its own foundation. The deep like a garment is its clothing. For just as the abode of fires is only in fire, of waters only in water, of air only in air, so the place of the earth is only in itself, with everything else enclosing it. Nature contains it and denies it any place to fall. Situated in the centre or pivot of the world, the earth, being the heaviest, holds the lowest and central place in creation, since water, air, and first as it were by the levity of their nature and likewise by their situation rank above it.[14]

And, finally, he says:

46. That the Earth Is Like a Globe

We say 'the sphere' of the earth, not because it has the shape of a perfect sphere, in view of so great a disparity of mountains and plains, but because, if all of its perpendicular lines were enclosed within a circumference, it would make the figure of a perfect sphere.[15]

We have dwelt on the quotations from *De natura rerum* with the conviction that this work represents (schematically) what was believed about the physical world at the beginning of the eighth century and, as we shall see, even beyond. One can say that the sphericity of the Earth (with the posthumous endorsement of Pliny!) was, at the moment, accepted, but the question of the celestial waters still remained. In fact, the revealed truth, which could not be disputed, was the biblical account of creation.

---

[13]Bede, *De Natura rerum, caput* VIII: *De aquis coelestibus. Aquas, super firmamentum positas, coelis quidem spiritualibus humiliores, sed tamen omni creatura corporali superiores, quidam ad inundationem diluvii servatas, alii vero rectius ad ignem siderum temperandum suspensas adfirmant* (Eng. trans. Calvin B. Kendall and Faith Wallis, op. cit., p. 77).

[14]Bede, *De Natura rerum, caput* XLV: *Terrae positio. Terra fundata est super stabilitatem suam, abyssus sicut pallium amictus eius. Sicut enim ignium sedes non est nisi in ignibus, aquarum nisi in aquis, spiritus nisi in spiritu, sic et terrae cohaerentibus cunctis nisi in se locus non est, natura cohibente, et quo cadat negante. Quae in centro vel cardine mundi sita, humillimum in creaturis, ac medium, tamquam gravissima, locum tenet, cum aqua, aer, et ignis ut levitate naturae, ita et situ se ad altiora praeveniant* (Eng. trans. Calvin B. Kendall and Faith Wallis, op. cit., p. 97).

[15]*De Natura rerum,* Caput XLVI: *Terram globo similem. Orbem terrae dicimus, non quod absoluti orbis sit forma, in tanta montium camporumque disparitate, sed cuius amplexus, si cuncta linearum comprehendantur ambitu, figuram absoluti orbis efficiat* (Eng. trans. Calvin B. Kendall and Faith Wallis, op. cit., p. 97).

3.3 The Early Scholastics

This section refers to the beginning of the period (which will last until the fourteenth century), referred to overall as "of the Scholastics", which can be subvdivided into four subperiods, the first of which is of the Pre-Scholastics.[16] One can say that this subperiod corresponds approximately to the Carolingian renaissance, which goes from the end of the eighth century to the end of the ninth. The eighth and ninth centuries were characterized by the concentration of the surviving forces of culture in the great empires of the West: the Arabian Empire and the Carolingian Empire, both of which made an intellectual revival possible.

As regards Charlemagne, induced by the need to organize the administration of his empire, he summoned to his side several learned people to act as government officials and manage the various branches of the administration. Futher, he founded the Palatine School to train this new ruling class, and in 781 called from England to direct it Alcuin, who had been trained in the episcopal school of York. Charlemagne also entrusted Alcuin with the task of correcting the Vulgate, since he wanted the same version of it to be uniformly diffused in all the regions of the empire.

In fact, in order to make the administration of the empire uniform to his instructions, Charlemagne also attended to the liturgy of the Church, introducing Gregorian chants and the Benedictine rule (in a new accurate version, the same in all monasteries). Also part of this project of sociocultural hegemony, by means of Alcuin (who in the meantime had become bishop of Tours), was the invention of the Caroline minuscule. This new style of handwriting, much more legible than earlier medieval handwritings, favoured the diffusion of manuscripts and thus the development of culture. The ability to know and summon intellectuals of various nationalities to contribute to the organization of the administration of the empire is typical of the Carolingian Empire, as we shall also see in the case of John Scotus Eriugena.

Generally one can say that the period of the Scholastics is characterized by the problem of the relation between reason and faith (*fides et ratio*), that is, between the truth the man can attain by means of his natural powers and that which is revealed to him from on high and imposed by hierarchies. The division into subperiods we have mentioned above can be said to be correlated to the different ways of solving that problem. In the course of this section we shall inspect the contributions made by various Scholastics philosophers to the solution of the problem we are dealing with. Of course, since our task is restricted to identifying the opinions expressed with regard to the shape and the nature of the Earth (and, in some cases, its position in the universe), we shall be obliged to proceed in a peculiar way (that is, the philosophers we shall mention are not necessarily the most important); it will, however, always be noted when the relation (sometimes conflictual) between reason and faith is involved.

---

[16] Nicola Abbagnano: *Storia della filosofia*, 2 vols. (Torino: UTET, 1993).

It is important to point out that in the period of the Carolingian renaissance (precisely with the Pre-Scholastics) a different way of dealing with philosophical thought begins. No longer is the philosophical thought of the Greeks simply recovered in order to transfer it to the centuries to come, but instead a new thinking is worked out, one which must find an agreement between reason and faith. The continuous reference to the philosophers of the past has a merely instrumental value; their thought is used as a tool in service of the development of the new way of thinking.

The Carolingian renaissance came to an end with the last successor of Charlemagne, Charles the Bald (823–876), who called (this time from Ireland) the monk John Scotus Eriugena to direct the Palatine School. John Scotus was also given the job of translating from the Greek the work of Pseudo-Dionysius the Areopagite, which Louis the Pious (778–840) had received as a gift from the emperor of Byzantium. Scotus was learned in Greek philosophy, which he was able to read in the original language, and he would become the most important philosopher of this period, destined to influence the subsequent developments of the Scholastics. About him, Nicola Abbagnano says:

> Unexpectedly in the first half of the ninth century appears the great figure of John Scotus. In the cultural and speculative poverty of his time, this man, blessed with an extremely free spirit, an exceptional speculative intelligence and a broad Graeco-Latin erudition, appears as a miracle.[17]

The most important work of John Scotus is *De divisione naturae* (or *Periphyseon*), in five books. It is written in the form of a dialogue between *magister* and *discipulus* and is the first great philosophical writing of the Middle Ages. The leading problem is the search for an intrinsic agreement between reason and faith: between the truth attained by the free research and that revealed to man by the authority of the sacred writings and of the inspired writers. He holds that the authority of the Holy Scriptures is indispensable to men to reach the truth, but also that it must not deter him from that of which the right reason persuades him.

The work *De divisione naturae* is of pure Platonic origin and John Scotus divides nature into four parts:

> Master: It is my opinion that once four specific differences are singled out, nature can be divided into four kinds, the first of which is that which creates and is not created, the second which is created and creates, the third which is created and does not create, the fourth which neither creates nor is created. Indeed, these four are opposed alternatively two by two. In fact the third is opposed to the first, and the fourth to the second; but the fourth is placed among the impossible things whose being is not possible.[18]

---

[17]N. Abbagnano, *Storia della filosofia*, op. cit. II, Chap. 11 (our Eng. trans.).

[18]John Scotus, *De divisione naturae* I, 1: *Magister: Videtur mihi divisio naturae per quattuor differentias quattuor species recipere, quarum prima est in eam quae creat et non creatur, secunda in eam quae et creatur et creat, tertia in eam quae creatur et non creat, quarta quae nec creat nec creatur. Harum vero quattuor binae sibi invicem opponuntur. Nam tertia opponitur primae, quarta vero secundae; sed quarta inter impossibilia ponitur cuius esse est non posse esse* (our Eng. trans.).

As we see, the third part is the world itself, the universality of the sensible and non-sensible things which derive from the primordial causes due to the distributive and multiplicative action of the Holy Spirit.

Besides dealing with the world when discussing the interpretation of the verses of the Genesis, John Scotus was also interested in scientific-allegorical divulgation, which induced him to write a commentary on *De Nuptiis Philologiae et Mercurii* of Marcianus Capella: *Adnotationes in Marcianum Capellam*. This interest in the cosmography of Greek origin, tied together with his conception on the role of reason, led John Scotus to be the first Christian thinker of the Middle Ages (anticipating the Scolastics of the twelfth century) to hold the impossibility of the celestial waters. Let us see, in short, what his thinking is, beginning from the first book of *De divisione naturae*. The Earth is the centre of the universe:

> Master: In fact, all this world which appears to our senses moves with a continuous motion around its axis; that is, around the Earth, around which as a fixed centre other three elements, Water, Air, Fire unceasingly rotate. And in the same way, with a ceaseless, invisible motion, the universal bodies, and I mean the four elements, come together and form the particular bodies of the individual things. Once these are disgregated, (the elements) come back from the particular to universal bodies, always remaining immutable, as a fixed centre of simple things, because of their natural essence which cannot change, neither increasing nor lessening.[19]

Regarding the celestial waters, first of all John Scotus expounds the various opinions and interpretations of the Fathers of the Church who preceded him. He then proceeds to express his own thought which, so to speak, "dematerializes" the *aquae superiores*:

> Master: So I think that every created thing falls within one of these subdivisions: or it is wholly Body, or wholly Spirit, or something intermediate between them; since it is neither wholly body, nor wholly spirit but something of intermediate, for a reason of compromise between the extremes it proportionally receives in itself something from the spiritual nature as from the upper extreme, and something from the other nature which is wholly corporeal. For this reason, it exists by itself although sharing in the nature of its extremes. Therefore, anyone who has observed carefully can understand that this world consists of a threefold articulation, since, if it is considered according to the reasons for which it is constituted and substantially exists, it is not only known as spiritual but even wholly as spirit. In fact none of those who correctly philosophize has denied that the reasons of the bodily nature are spiritual, or better, are the spirit itself. But if instead one looks downward, at the lowest parts of the world, that is, at all these bodies made up of the universal elements, chiefly at those terrestrial and aqueous which are attached to generation and corruption, he will not find anything other than body and corporeity. Yet, anyone who turns to look at the nature of the simple elements, will find with extreme clarity a certain mean proportion owing to

[19]John Scotus, *De divisione naturae* I, XXXII: *Etenim totus iste mundus sensibus apparens assiduo motu circa suum cardinem volvitur; circa Terram dico, circa quam veluti quoddam centrum caetera tria elementa, Aqua videlicet, Aer, Ignis, incessabili rotatu volvuntur, et ita invisibili sine ulla intermissione universalia corpora, quatuor dico elementa in se invicem coeuntia singularum rerum propria corpora conficiunt. Quae resoluta iterum ex proprietatibus in universalitates recurrunt; manente semper immutabiliter quasi quodam centro singularum rerum propria naturalique essentia, quae nec moveri, nec augeri, nec minui potest* (our Eng. trans.).

which they neither are wolly bodily, notwithstanding the natural bodies survive their corruption thanks to coitus, nor are completely devoid of bodily nature, since from them all bodies derive and in them again come back. And again, according to what is said about the reciprocal correspondence with the upper extreme, they neither are wholly spirit since they are not completely released from the corporeity, nor wholly non-spirit since they get from wholly spiritual reasons the occasions of their subsistence. Thus, not unreasonably we said that this world possesses extremes deeply separate from each other and intermediate entities in which the consonant harmony of this universe gathers. Let us say then that the lower parts of this world are the lower waters. And this not without reason, since all which comes up in this world grows and feeds on water; in fact, if one removes the moisture from the bodies, they swiftly run out and diminish and reduce to nearly nothing. In fact the pagan philosophers maintain that also the hottest and burning celestial bodies feed on the aqueous moisture and this is not even denied by the interpreters of the Holy Scriptures. The reason teaches (*ratio edocet*) that the spiritual causes of all which is visible are called upper (celestial) waters. In fact, from these all elements, both simple or compound, derive as from great sources and are organized in conformity with intelligible virtues stimulated by them.[20]

[20]John Scotus, *De divisione naturae* III, XXVI, pp. 254–255: *Totius itaque conditae naturae trinam dimensionem esse arbitror; omne enim quod creatur, est vel omnino Corpus, vel omnino Spiritus, aut aliquod medium: quod nec est omnino corpus, nec omnino Spiritus, sed quadam medietatis, et extremitatum ratione, et Spirituali omnino natura, veluti ex una extremitate et superiori et ex altera, hoc est, ex omnino corporea proportionaliter in se recipit, unde proprie, et connaturaliter extremitatibus suis subsistit; proinde siquis attentus inspexerit in hac ternaria proportionalitate hunc mundum constitutum intelliget; siquidem in quantum in rationibus suis, in quibus essentialiter et constitutus est, et essentialiter subsistit consideratur, non solum spiritualis, verum etiam omnino spiritus cognoscitur, nemo enim recte Philosophantium rationes corporeae naturae spirituales, imo etiam spiritus esse negarit, dum vero extremae ipsius deorsum versus inspiciuntur partes, hoc est, omnia ista Corpora ex Catholicis elementis composita, maxime etiam terrena, et aquatica, quae et generationi, et corruptioni obnoxia sunt, nihil aliud in iis reperitur praeter corpus, et omnino corporeum: at si quis simplicium elementorum naturam intueatur, luce clarius quandam proportionabilem medietatem inveniet, qua nec omnino corpus sunt, quoniam eorum corruptione naturalia corpora subsistunt, et coitu: nec omnino corporeae naturae expertia, dum ab eis omnia corpora profluant, et in ea iterum resolvantur, et iterum alterum alteri superiori quidem extremitate comparata, nec omnino spiritus sunt, quoniam non omnino corporea extremitate absoluta, nec omnino non spiritus, quoniam ex rationibus omnino spiritualibus subsistentiae suae occasiones accipiunt; non irrationabiliter itaque diximus, hunc mundum extremitates qnasdam a se invicem penitus discretas, et medietates, in quibus universitatis concors harmonia conjungitur, possidere. Ponamus igitur inferiores hujus mundi partes, veluti inferiores aquas: nec immerito, dum totum quod in hoc mundo nascitur, humore crescit, atque nutritur, humida siquidem qualitate corporibus sublata, absque mora tabescunt, et decrescunt, et pene ad nihilum rediguntur: nam et caelestia corpora ferventissima, et ignea humida aquarum natura nutriri sapientes mundi affirmant quod nec sacrae scrtpturae expositores negant. Spirituales vero omnium visibilium rationes superiorum aquarum nomine appellari ratio edocet. Ex ipsis enim omnia elementa, sive simplicia; sive composita, veluti ex quibusdam magnis fontibus defluunt, indeque intelligibili quadam virtute rigata administrantur* (our Eng. trans.).

Finally he concludes:

> There are some who think that over the firmament, that is, over the constellations of the stars, there are very tenuous waters, but the weight and the order of the elements refute them.[21]

Formerly he had been ironic (when speaking of the firmament):

> …thus, each one has explained, as to his way of thinking, the meaning of that name: then some, since it carries the upper (celestial) waters (as if over the firmament there should be real waters)….[22]

We add that, further on in Book XXXIV of this work, John reports Erathostenes' measurement of the terrestrial circumpherence, dwelling also on the description of the scaphe. The measure he quotes is that of 252,000 stades. It is not clear if his narration derives from a direct reading of a Greek work, or if it is only a new proposal of Martianus' text, which is perhaps the most likely hypothesis.

In the tenth century, the general (political-social) conditions worsened after the disintegration of the Carolingian Empire and a cultural decline also followed. There was a short revival with the Ottonian Empire (962–1024). An unusual figure of scholar belongs to this period, who also was an influential abbot destined to become bishop and then pope. This was Gerbert of Aurillac. He was educated in the school of Aurillac, then went as a teacher to Rheims (in 972), then to Bobbio (in 982) and then back again to Rheims (in 991) as archbishop, finally becoming pope (in 999) with the name of Sylvester II. He died in 1003.

He spent a certain period of time in Catalan Spain, at that time seat of circles where Hellenistic science translated into Arabic by Muslim scholars was widespread. Historians attribute to this stay the new elements of astronomy and mathematics that Gerbert shows himself to possess, as can be seen in his letters,[23] in which he demonstrates his interest in a wide range of subjects, as well as in his extant works.[24] The fact of being in the vanguard (for his times) of scientific knowledge also made him the target of the envy of his contemporaries, and considerable falsehoods were spread about him (he was even accused of necromancy), making him a character of legend.[25] In spite of it all, he became pope.

Even if in his works there is no specific treatment expounding the cosmology and the placement of the Earth in the universe, we mention him here because his knowledge of the Hellenistic cosmology (credited to his time in Spain) and his

---

[21]John Scotus, *De divisione naturae* III, XXVII: *Sunt qui tenuissimas aquas supra firmamentum, i.e. super choros siderum esse putant, sed eos refellit et ratio ponderum, et ordo elementorum* (our Eng. trans.).

[22]John Scotus, *De divisione naturae* III, XXVI: *...quare autem tali nomine vocatur, prout unicuique visum est, ita exposuit; alii quidem praeter sustentationem aquarum superiorum, veluti supra illud corporales aquae sint...* (our Eng. trans.).

[23]Harriet Pratt Lattin, trans.: *Letters of Gerbert* (Columbia University Press, 1961).

[24]N. M. Bubnov: *Gerberti opera mathematica* (Berlin, 1899).

[25]A. Graf: *Miti, leggende e superstizioni del medio evo* (Torino: Loescher, 1892–1893; rpt. Bruno Mondadori, 2002). The relevant essay is at pp. 201–255.

reputation in the schools of the monasteries give us elements of confirmation that the Ptolemaic two-sphere model was transmitted in the schools of the eleventh century.

In this regard, it is also important to point out that Gerbert, in his work *Geometria de ponderibus et mensuris*, talks about the measurement of the circumpherence of the Earth by Eratosthenes and quotes the measure of 252,000 stades. Also here, as in the case of John Scotus, it is not clear if Gerbert goes back to Martianus or if, during his stay in Spain in a circle strongly permeated by the Arabic culture, had gained access to some Greek text.

We have been unable to find in Gerbert's writings references to the "celestial waters" and the discussions related to them, but this does not mean that the subject had been set aside. As we shall see, in the subsequent centuries, in the cathedral schools (Chartres, etc.) the subject became once again of fundamental importance.

3.4 The Problem of the Supercelestial Waters in the Scholastics of the Twelfth Century

In the twelfth century, there was a general improvement in the standard of living (in fact, some historians talk of the "renaissance" of the twelfth century) and an economic recovery which also influenced the development of the schools (at that time in the cathedrals). As we have seen, the influence of Neoplatonism, and mainly that of the authority of Plato's *Timaeus* in this period, meant that the spherical shape of the Earth, the two-sphere model of the universe, and the four elements were accepted without discussions. The times of Lactantius (whose positions, to tell the truth, were never made official by the Church) were far in the past and the problem of antipodes had not yet become topical again. But the problem of the supercelestial waters remained: it was the wall against which those who wished to somehow express in consistent terms the obscure verses of Genesis banged their head.

In this period the exegesis of the Bible was only practised within the religious milieu; a secular intellectuality did not exist and the question was not debated between opposing sides. But the problem nevertheless posed itself: Scholastics philosophers were also obliged to look for an explanation of a statement in the Holy Bible which, in its literal meaning, contradicts the physical reality. It was no longer possible to set aside the problem by deferring all to the will of the Creator (who can make water to float on air); a solution had to be found which did not overly mortify the physical reality.

Whereas before opposition to the idea of the spherical shape of the Earth had not been based on religion, now the problem was posed directly by the sacred scriptures. Thus began the conflict between the "truth" of the Bible and physical reality in accordance with science. This is the reason why we deal with the problem of the supercelestial waters, which has nothing to do with the shape of the Earth, but obviously with its position in the universe.

As we shall have opportunity to see, the problem is substantially transformed, so to speak, into a non-problem in the ambit of the Scholastics, and thus no conflict between science and faith is left on the field. We shall shortly inspect some of the solutions which, in the twelfth century, in one way or another refer to Platonic philosophy by associating Plato's "world soul" to the Holy Spirit.

3.4.1 Abelard

Peter Abelard (1079–1142) is undoubtedly the personality who characterizes the philosophical debate in the first half of the twelfth century. Though nowadays he is also known to a large audience of non-philosophers owing to his dramatic love story with Heloise, in his own time he dominated the scene of the philosophical and theological disputations thanks to his profound learning and rigorous logic. The centre of his personality was the need to research: he wanted to solve on rational grounds every revealed truth, and to face with the tools of dialectic any problem in order to trace it back to an effective human comprehension.

His main work, to which we shall refer here, is the *Expositio in Hexaemeron*. In it, he devotes himself, as Augustine had done before him, to explaining and clarifying the obscure expressions and conflicting statements of the sacred text. However, in comparison with Augustine, he takes a step forward, striving to give a philosophical (physical) explanation of nature, strongly influenced by the Platonic vision. Let us see an excerpt of Abelard's commentary:

> Some maintain that these waters are retained for the Deluge, others, more correctly, say they have been set in order to mitigate the fire of the stars. The blessed Augustine, leaving out all these opinions about the upper waters (that is, if they are iced or not, or if they are of some avail), says: "Certainly only the Creator knows of what type these waters are and to what employment they are reserved, what is not in doubt is their existence because of the witness of the Scripture". Therefore, since so great a doctor remained in some doubt, an assertion on our part would appear very arrogant. Yet, since some said they were arranged and preserved for the inundation of the Deluge so that they covered the earth falling abundantly, this is an absolutely foolish opinion. …
>
> What utility then that suspension of the waters can have, I think it is very difficult to decide since not even to the Holy Fathers it appeared established with a non-questionable reasoning. However, the opinion which appears to us to be the most probable (*probabilior …. opinio*) is that according to which they have been constituted above all for mitigating the heat of the upper fire, so that its fieriness does not completely attract the clouds or the lower waters, the strength of the fire being such as to naturally attract the humidity: …
>
> And it must be pointed out that where we say "Let the firmament be in the middle of the waters", the Hebrews have "let there be some space under the waters"; that is, an interval by which they are separated between themselves and thanks to which they never will come to touch.[26]

[26]Abelard: *Expositio in Hexaemeron*—P. L, Tome CLXXVIII, cols. 743–744: *Quidam ad inundationem diluvii reservatas, alii vero rectius ad ignem siderum temperandum suspensas affirmant.*

In other words, the procedure adopted by Abelard means that, if it is not acceptable to defer the question to divine omnipotence, the solution, even if only probable, must be found by using the tools that the man has at his disposal: reason and knowledge of nature.

Some pages later (col. 748) Abelard also deals with the waters that wash the Earth, that is, the relation of the Earth with the seas:

> Just as when a globe is plunged in water in such a way that a part of it emerges, so the globe of the earth lies in the waters so that one part of it is in contact with the sea....[27]

As one can see, this is not the Aristotelian model of the four elements as four concentric spheres.

3.4.2 The School of Chartres

As we have already said, in the twelfth century the schools passed from the monasteries to the episcopal seats in the cathedrals. The school which flourished at the Cathedral of the French town Chartres is the most renowned and several philosophers and theologians are dated back to it. In the School of Chartres the culture that was taught was founded in the arts of the trivium (grammar, rhetoric, dialectic) and of the quadrivium (arithmetic, geometry, astronomy, music) in a harmonic synthesis. There both the Latin classics (Virgil, Ovid, Horace, Lucan, Cicero, Seneca) and the scientific thought based on the encyclopaedias of Isidore, Bede and Pliny the Elder. Recent acquisitions derived from the translations from the Arabic were also studied. Above all, the *Timaeus* was studied and considered the indispensable tool for interpreting Genesis.

Beatus vero Augustinus istas opiniones praetermittens de aquis illis superioribus, utrum videlicet glaciales sint vel non, vel quas in se habeant utilitates, ait; "Sane quales ibi aquae sint, quosve ad usus reservatae, Conditor ipse noverit, esse tamen eas ibi Scriptura testante nulli dubium est". Quod ergo tantus doctor quasi dubium sibi reliquit, affirmare nobis arrogantissimum videtur. Quod vero nonnulli opinati sunt eas ibi constitutas et reservatas ad inundationem diluvii, ut inde scilicet labentes abundantia sui terram cooperirent, frivolum omnino deprehenditur. ...

Quid ergo suspensio illa aquarum utilitatis habeat, quod nec a sanctis certa sententia definitum est, difficillimum disseri arbitror. llla tamen nobis probabilior videtur opinio, ut ob hoc maxime ad calorem temperandum saperioris ignis constituerentur, ne fervor ille superior vel nubes ipsas vel aquas inferiores omnino attraheret, cum sit vis ignis naturaliter attractiva humoris. ...

Et notandum quod ubi nos dicimus Fiat firmamentum in medio aquarum Hebraei babent: sit extensio infra aquas; hoc est intervallum quo ipsae ab invicem in perpetuum separentur ne se ulterius contingant (our Eng. trans.).

[27] *Quasi enim aliquis globus ita in aqua constituatur, ut una pars eius superemineat, ita ille globus terrae in aquis insedit, not ex una parte cum mare contingeret...*"; see W. G. L. Randles, "Classical Models of World Geography and Their Transformation Following the Discovery of America", p. 22, in *The Classical Tradition and the Americas*, W. Haase and M. Reinhold (eds.) (W. de Gruyter, 1994).

A reading of the Bible was now more carefully tied to the natural phenomena. It could be said that it was a reading *secundum physicam*, but the term *physica* must not deceive us. It did not have the meaning which we attribute to it nowadays; it was simply a more rational reading than that of the past, with a greater attention to nature. For our purposes, we shall limit ourselves to quoting three of the most representative philosophers of the school of Chartres still with the aim of ascertaining which was the most reliable opinion regarding the verses of Genesis on the Earth and the supercelestial waters.

3.4.2.1 Thierry (Theodoric) of Chartres

Thierry, originally from Brittany, arrived to the Cathedral of Chartres together with his brother Bernard. According to Abelard, he was already *magister* there in 1121. It is likely that he lived in Paris between 1124 and 1141. It is known that when he was elderly, he retired to a monastery. The exact year of his death is unknown, but it is in any case later than 1156. These are the only certain biographical notes. Among his main works there is *Tractatus de sex dierum operibus*, which begins with the phrase "*secundum physicam*" we have already mentioned.

He says he will explain the first part of Genesis "in accordance with natural science [physicam] and the literal sense [of the text]",[28] neglecting the allegorical and moral interpretations already dealt with by the holy interpreters. In his treatise, Thierry "explains" the formation of the world rationally through the reading and interpretation of the early chapters of Genesis. We report the excerpt in which he deals with the supercelestial waters:

> [Day 1. Genesis 1:1–5] And once the air was illuminated by the power of the higher element [fire], it followed naturally that, by illuminating the air, fire heated the third element, water, and by heating it suspended it as vapour above the air. For it is the nature of heat to divide water into very tiny drops and raise those tiny drops above the air by the power of its motion, as is seen in steam in a boiling pot and as appears in the clouds in the sky. For the clouds or steam are nothing other than masses of very tiny drops of water raised into the air by the power of heat. If the power of heat becomes stronger, the whole mass changes into pure air. If, however, it becomes weaker, then those very tiny drops, rushing together, make bigger drops and cause rain. But if those tiny drops are compressed by wind, they cause snow, and if they become large, they make hail. [Day 2 Genesis 1:6–8] Therefore the huge mass of moving water, which in the beginning doubtless reached to the region of the moon, was suspended by heat above the height of the ether, so that immediately, during the second rotation of the heaven, it happened that the second element, air, was in the middle between the moving water and the water suspended as vapor. This is what the author refers to when he says, *And he placed the firmament in the middle of the waters* (Genesis 1:6). Then it was fitting to call the air the "firmament," since it firmly supported the water above it and restrained the water below it, making sure that they did not cross over into one another. Or perhaps air is called the "firmament" rather because by its lightness it firmly restrains the earth from every direction and forms it into a hard ball. For

[28] Thierry of Chartres, *Tractatus de sex dierum operibus*, 1: *secundum physicam et ad litteram* (eng. trans. Katharine Park).

> there is a reciprocal relationship between the hardness of the earth and the lightness of the air: the hardness of the earth comes from the constriction of the light air, and the lightness and mobility of the air has its substance from the fact that it rests on the stability of the earth.[29]

3.4.2.2 William of Conches

William was born about the end of the eleventh century in a little town of Normandy. He may have begun his teaching activity between 1120 and 1125; he himself was a pupil of Bernard of Chartres. Among his works, we are particularly interested in the early work *Philosophia mundi* and in *Dragmaticon philosophiae*, written when he was middle-aged. The latter is in form of a dialogue between the "Duke" (The Duke of Normandy) and the "Philosopher"; in the prologue he says he will repudiate part of the statements made in *Philosophiae mundi*:

> There is, however, a little book of ours on the same subject, entitled Philosophia; it is quite imperfect, as it was composed in our imperfect youth. In that booklet truths are interspersed with falsehoods and many point that out to have been made were omitted.[30]

The statements he repudiates concern theological subjects (for instance, Father, Son, and Holy Spirit and the creation of the woman from the mud instead of the

---

[29]Thierry of Chartres: *Tractatus de sex dierum operibus*, 7–8: *Aere uero ex superioris elementi uirtute illuminato, consequebatur naturaliter ut, ipsius aeris illuminatione mediante, calefaceret ignis tercium elementum i.e. aquam et calefaciendo suspenderet uaporaliter super aera. Est enim natura caloris aquam in minutissimas guttas diuidere et eas minutas uirtute sui motus super aera eleuare sicut in fumo caldarii apparet: sicut etiam in nubibus celi manifestum est. Nubes enim siue fumus nichil est aliud quam guttarum aque minutissimarum congeries per uirtutem caloris in aera eleuata. Sed si uirtus caloris uehementior fuerit, tota illa congeries in purum aera transit: si autem debilior tunc nimirum gutte ille minutissime semet inuicem incurrentes grossiores guttas faciunt: et inde pluuia. Quod si minute ille gutte uento constricte fuerint inde nix: si uero grosse inde grando. Magnitudo igitur aquarum labilium que nimirum usque ad regionem lune in principio ascendebat ita per calorem super summum etheris suspensa est ut statim in secunda celi conuersione ita contingeret quod secundum elementum i.e. aer esset medium inter aquam labilem et aquam uaporaliter suspensam. Et hoc est quod dicit auctor: et posuit, "firmamentum in medio aquarum". Et tunc aer aptus fuit ut, "firmamentum" appellaretur quasi firme sustinens superiorem aquam et inferiorem continens: utramque ab altera intransgressibiliter determinans. Uel potius, "firmamentum" dicitur aer eo quod terram leuitate sua ex omni parte firme coherceat et in hanc duriciam conglobet. Est enim ista reciprocatio inter terre duriciem et aeris leuitatem ut durities terre ex circumstrictione leuis aeris proueniat: leuitas uero aeris atque mobilitas ex eo quod terre stabilitati innititur habet substantiam* (Eng. trans. Katharine Park).

[30]William of Conches, *Dragmaticon Philosophae* I, 1: *Est tamen de eadem materia libellus noster qui PHILOSOPHIA inscribitur, quem in iuventute nostra imperfectum, utpote imperfecti, composuimus, in quo veris falsa admiscuimus multaque necessaria praetermisimus. Est igitur nostrum consilium, quae in eo vera sunt hic apponere, falsa damnare, praetermissa supplere* (Latin text from *Guillermi De Conchis Opera Omia*. Tomus I: Pragmaticon Philosophiae, ed. I. Ronca (Turnhout: Brepols, 1997), p. 7; Eng. trans. Italo Ronca and Matthew Curr).

Adam's rib). What he does not repudiate is the statement that the supercelestial waters do not exist.

We report first the relevant excerpt from *Philosophiae mundi*:

> Some say that over the ether there are congealed waters, taut like a skin, which appear to our eyes, and over them waters whose existence is confirmed by the authority of the Holy Scripture, which says: *He put the firmament in the midst of the waters*, and again, *He divided the waters which are under the firmament from those which are over the firmament* (Gen 1). But, since this is against reason, we show why things cannot be in this way and then how the aforesaid sentences of the Holy Scripture must be understood. If in that place there are congealed waters, then there is something weighty in itself, but the natural place of the weighty things is the earth. Moreover, if in that place there are congealed waters will they be joined to the fire or not? If they are joined to the fire, it being warm and dry, and the congealed water cold and humid, the opposites would be joined without any intermediary; then there would never be concordance but the repugnance of the opposites. Even more: if the congealed water is joined with the fire whether it will be dissolved by the fire or it will extinguish the fire. So, if the fire and the firmament remain, where are the congealed waters, joined with the fire; if they are not joined, is there something between them? But what? An element? But no element made by elements and then visible, can stay over the fire. Thus, not being there anything visible, it follows that there are no congealed waters.[31]

And, for comparison, here is the excerpt from *Dragmaticon*:

> DUKE: Bede says that what appears to us like a taut skin are congealed waters.
>
> PHILOSOPHER: In those things which concern the Catholic faith or the moral education it is not permissible to contradict Bede or any other of the saint Fathers. However, in those things which concern physics, if they are wrong in something, it is permissible not to agree. In fact, even though greater than us, they were nevertheless men.
>
> DUKE: When an inferior contradicts one who is superior to him, he is obliged to indicate the reason why the things must be otherwise. If, then, you want to be believed in this regard let the reason, true or verisimilar, be shown by which the things cannot be as Bede says.
>
> PHILOSOPHER: The water by its nature is weighty; if congealed, since it is in this case closer to the nature of the earth, it would be even weightier and the motion peculiar of what is weighty is toward the center. Then, if there were congealed waters over the ether, because of natural gravity they would descend down to the deepest place.

---

[31]William of Conches, *De Philosophia mundi*, II, II: *Dicunt enim quidam super aethera esse aquas congelatas, quae in modum pellis extensae. occurrunt oculis nostris, super quas aquae sunt confirmatae, hac auctoritate divinae paginae, quae ait:* Posuit firmamentum in medio aquarum *et iterum: Divisit aquas, quae sunt sub firmamento ab* his quae sunt super firmamentum *(Gen. 1). Sed, quoniam istud contra rationem est, quare sic esse non possit ostendamus, et qualiter divina Scriptura in supradictis intelligenda sit. Si ibi sunt aquae congelatae, ergo aliquid ponderosum et grave? Sed primus locus ponderosorum et gravium est terra. Si iterum ibi sunt aquae congelatae, vel igni conjunctae, vel non? si igni conjunctae sunt, cum ignis calidus et siccus sit, aqua congelata et humida, contrarium sine medio suo contrario conjunctum est. nunquam ergo ibi concordia, sed contrariorum repugnantia? Amplius: Si aqua congelala conjuncta est igni vel dissolvetur ab igne, vel exstinguit ignem: cum ergo ignis et firmamentum remanent, ubi sunt aquae congelatae, conjunctae igni, si conjunctae non sunt, aliquid inter eas et ignem est? Sed quid? elementum, sed nullum superius igne factum est ex elementis, visibile. ergo. Unde igitur non videtur. Restat ibi non esse aquas congelatas* (Latin text from *Patrologia Latina*, vol. CLXXII, cols. 57–58, where it is attributed to Honorius of Autun; our Eng. trans.).

DUKE: We see that heavy bodies are supported by the vault of the arch and never fall: thereby these congealed waters, being supported like a vault cannot fall.

PHILOSOPHER: If the arch which supports the vault had not basements leaning on something solid, inevitably would fall at once. Hence, this arch you imagine rests its basement on something solid, but there is nothing solid except the earth. Therefore, it is necessary that those basements are fixed to the earth: and this is worse then a stupid shaft of wit.[32]

Previously, the Duke had asked a question out of the ordinary (at least in that time). Here is the question and the reply of the philosopher:

DUKE: All heavy things, conforming themselves to the nature of the earth, by their nature tend downwards. Then I want you to say, if the earth was bored all the way to the opposite surface, and a stone dropped into that hole, what would happen?

PHILOSOPHER: Even if, with respect to the other three elements, the earth is said to be the lowest one, however its middle part is lower than the other of its parts. Hence, if this stone was dropped from the surface into the hole, it would descend to the middle of the earth and if by an accidental solicitation it went beyond [the middle], it again would return to the middle by natural motion; and if it went beyond [again], it would exceed a little and would come back to the centre at once, but going less far beyond.

And so, over and over returning but always going less and less far beyond, it would settle at the bottom and there would remain.[33]

---

[32]William of Conches, *Dragmaticon* III, 2: *DVX: Beda dicit aquas congelatas esse quae ad modum pellis extensae nobis apparent.*

PHILOSOPHVS: In eis quae ad fidem catholicam uel ad institutionem morum pertinent, non est fas Bedae uel alicui alii sanctorum patrum contradicere. In eis tamen quae ad physicam pertinent, si in aliquo errant, licet diuersum adfirmare. Etsi enim maiores nobis, homines tamen fuere.

DVX:Quotiens minor maiori contradicit, rationem, quare aliter esse oporteat, oportet inducat. Si uis igitur ut super hoc tibi credatur, quare sic esse non possit ut Beda dicit, uel uera uel uerisimilis ratio adducatur.

PHILOSOPHVS: Omnis aqua naturaliter est ponderosa; congelata, quia plus accedit ad naturam terrae, est ponderosior. motus proprius ponderum est ad centrum. Si igitur aquae congelatae super aethera essent, naturali grauitate ad ima descenderent.

DVX: Videmus grauia per fornicem arcus sursum teneri neque umquam descendere: unde aquae istae congelatae, quia in modum fornicis sunt suspensae, descendere non possunt.

PHILOSOPHVS: Arcus fornicis, nisi bases quae alicui solido innitantur habeat, necesse est statim corruat. Iste igitur tuus arcus, quem confingis, alicui solido basem figit, sed nichil est praeter terram solidum. Oportet igitur quod istae bases terrae inhaereant: quod uilitatem scurrilis ioci excedit (Latin text from *Guillermi De Conchis Opera Omia*, op. cit., pp. 58–59; our trans.).

[33]William of Conches, *Dragmaticon* II, 6*: DVX: Grauia omnia, naturam terrae sequentia, naturaliter ad imum tendunt. Volo igitur dicas, si terra esset forata usque ad alteram superficiem, lapisque in foramine illo demitteretur, quid esset?*

PHILOSOPHVS: Etsi terra respectu aliorum trium dicitur infima, media tamen illius pars inferior est aliis partibus eiusdem. Vnde, si hoc esset lapis a superiori superficie in foramen demissus, usque ad medium terrae descenderet; et si ultra ex accidentali impetu transiret, iterum naturali motu ad medium reuerteretur; quod si transiret, parum transiret, statimque ad medium rediret, minus tamen transeundo; et ita, frequenter ad imum reuertendo minusque et minus transeundo, imo adhaereret, ibique staret (Latin text from *Guillermi De Conchis Opera Omia*, op. cit., p. 53; our Eng. trans.).

The same question will be dealt with two centuries after by Galileo in his *Dialogue*, but with a different conclusion. As we see, the Philosopher concludes that the motion will be a damped oscillatory motion, whereas Galileo will maintain that it is a periodic oscillation.[34]

3.4.2.3 Bernard Silvester

As for so many others, for Bernard Silvester as well the biographical notes are almost null. He may have been a teacher in the Abbey of Tours between 1130 and 1140; he died before 1178. Among his works, we shall consider two in which there are statements regarding the Earth. His main work is *Cosmographia* (or *De mundi universitate libri II seu Megacosmus et Microcosmus*) which is in both prose and verses. The other work we shall consider is the *Commentary* to Martianus Capella. From the former, we take the excerpt in verses:

> I set it to my praise and glory, O Nature, that I have so well cultivated my coarse materials. I have brought form to creatures and yoked the elements by a harmony which has eleicted peace and trust. I have given a law to the stars, and order the planets always to pursue the same undeviating course I have curbed the sea with boundaries, lest the land be flooded, and the earth rests, fixed by its own weight, at the center of things. I have decreed that ethereal warmth should bring forth vegetation, and that moisture should sustain what this ethereal warmth has produced; that the earth, loving mother, should give birth to all things, and at thei dissolution receive them back again into her tranquil depths; that every creature should derive the seed and principle of its vitality from Endelechia, soul of the universe.[35]

And from the latter we take:

> Hence, supported by these authors, let us say that the supercelestial waters exist and it is due to them if Saturn is cold and humid, as is the Moon for the nearness of the lower elements, whereas Mars is hot for the nearness of the Sun. Indeed, what those think, that the looser elements cannot bear a denser substance, is not philosophical, in fact, we see that both wood and stones are borne by waters. And also the air close to the earth, even if lighter than waters, supports them when they are turned into steam. If then the higher waters are less dense than those turned into steam, why could they not be perpetually supported by the fire and the air, as those denser are lifted by the air from the ground? In fact, it is manifest that the dense clouds and the huge bodies of dragons or birds are supported by the air. And furthermore, the air closed in a bladder maintains the skin of the bladder stretched, even if lighter, and will be able of supporting a weight as great as it be, until it is kept closed. Thus,

---

[34]For this see, D. Boccaletti: *Galileo and the equations of motion* (Springer, 2016), p. 119.

[35]Bernard Silvester, *Cosmografia*: *In laudem titulosque,meos, Natura, repono/Tam bene materias excoluisse rudes/Induxi rebus formas, elementa ligavi/Concordem numero conciliante fidem. / Ascripsi legem stellis lussique planetas/ Indeclinatum currere semper iter. /Substrinxi mare limitibus, ne terra labaret: /In medio sedit pondere fixa suo. /Mandavi calor aethereus produceret herbas, /Quas calor aethereus parturit, humor alat. /Corpora cuncta creet tellus resolutaque rursus/Excipiat placido mater amica sinu. /Ex endelechia mundi res quaeque creata/Sementem vitae principiumque trahat* (Latin text from *Bernardi Silvestris De mundi universitate libri duo*, eds. Carl Sigmund Barach and Johann Wrobel (Innsbruck, 1876), p. 35; Eng. trans. Winthrop Wetherbee).

> a bubble of fire or air, closed in the density of the waters, notwithstanding its lightness, is not forbidden from supporting the waters, neither the water can fall around, until the fire or the air collapse somewhere, since a body cannot occupy the place of another if this has not drawn back. Indeed, the air and the fire, in order not to fly away, are compressed by the surrounding waters and have waters overlapping from everywere. And if those waters, thickened by the chilliness, become hard like a crystal, the more they are solid, so much keep closed the air and the fire so that they cannot escape and the more they are strongly supported. Indeed, it is not even necessary to support them since they are no more fluid.[36]

In the first excerpt, Bernard shows that he still believes in a model of a Earth encircled in the "great river Oceanus", and in the second that he does not argue *secundum physicam*.

3.4.3 Alexander Neckam

What we can say after having reported the three examples above is that even if in a general philosophical survey (which we, absolutely, have not drawn) it does exist a characteristic which identifies the Scolastics of the twelfth century, in the case of our particular problem that characteristic does not exist. Perhaps, as we have said above, it was no longer considered an important problem as in the preceding centuries, that is, one could have different ideas on the supercelestial waters without being considered heretical.

This attitude can be clearly grasped if we consider the writing of a personality typical of that age, although a very particular one at the beginning of his life. He died in the second decade of the thirteenth century, but the work we shall refer to

---

[36]Bernard Silvester, *The commentary on Martianus Capella's De Nuptius Philologiae et Mercurii*: *His ergo auctoribus adquiescentes, dicamus aquas supercelestes esse et ex eis Saturnum frigidum et humidum, sicut luna ex inferiorum elementorum vicinia. Mars vero ex vicinitate solis calet. Nam quod illi estimant rariora elementa corpulentiorem substantiam sustinere non posse, philosophicum non est. Nam et ligna et lapides ab aquis sustentari videmus. Vicinus quoque aer, quamvis aquis levior, eas tamen vaporaliter tractas sustinet. Si ergo his aquis vaporatis ille superiores rariores sunt, cur non ab igne et aere perhenniter sustentarentur, cum hee corpulentiores a solo aere suspendantur? Nam et densas nubes et ingentia draconum vel avium corpora ab aere sustentari manifestum est. Preterea aer clausus in vesica circumstantem pellem vesice undique suspendit, licet ea sit levior, et etiam quantamcumque molem sustinere poterit quamdiu ibi clausus tenebitur. Sic interior ignis et aeris globus in illa aquarum clausus corpulentia nequaquam sua levitate eas suspendere impeditur, nec usquam labi aqua circumfusa poterit, donec ei ignis vel aer in aliquam partem cedat, quoniam locum unius corporis nullatenus alterum occupare potest nisi illo prius cedente. Undique vero aer et ignis, ne devolare queant, circumstantibus aquis comprimuntur et undique superpositas habent aquas. Quod si aque ille glaciali constrictione quasi in cristallum indurare sunt, quanto magis solide, tanto magis inclusum aerem et ignem cohibent, ne aliquo abscedant et tanto fortius ab eis sustentantur; immo nec eas sustentari necesse est que iam fluide non sunt* (Latin text from Haijo Jan Westra (ed.): *The commentary on Martianus Capella's De Nuptius Philologiae et Mercurii attributed to Bernardus Silvestris* (Pontifical Instutute of Mediaeval Studies, 1986), pp. 187–188; our Eng. trans.).

was written still in the twelfth century. We quote again the work of Lynn Thorndike to introduce his biography:

> In the year 1157 an Englishwoman was nursing two babies. One was a foster child; the other, her own son. During the next fifty years these two boys were to become prominent in different fields. The fame of the one was to be unsurpassed on the battlefield and in the world of popular music and poetry. He was to become king of England, lord of half of France, foremost of knights and crusaders, and the idol of the troubadours. He was Richard, Coeur de Lion. The other, in different fields and a humbler fashion, was none the less also to attain prominence; he was to be clerk and monk instead of king and crusader, and to win fame in the domain of Latin learning rather than Provençal literature. This was Alexander Neckam.[37]

Neckam was born at St. Albans, taught for some years at Dunstable, was abbot of St. Albans (1183–1195), and later became a canon of Circencester, where he was made abbot in 1213 and died in 1217. He studied and taught in Paris and also visited Italy. He was a man of deep classical culture and also knew part of Aristotle's works through the recent translations of the twelfth century. He wrote works on various subjects as well as verses in Latin. Again quoting Lynn Thorndike: "Neckam gives us a glimpse of the learned world of his time as well of his education".[38]

We shall only deal with one of his works in particular: *De naturis rerum* (*The natures of things*), which "is not primarly a scientific or philosophical dissertation, as Alexander is careful to explain in the preface, but a vehicle for moral instruction".[39] As to the problem of the supercelestial waters, in Chap. XLIX of the second book of this work, Alexander says:

> One will object to what the prophet says: "God has consolidated the earth over the waters". From this there would seem that one can think that the waters are below the earth, whereas Alfraganus says that the sphere of the waters and the earth is only one. The interpreters of the Holy Scripture attribute the phrase of the prophet to the use of the everyday language, as it is usual to say that Paris is founded on the Seine. But the truth is that the Eden is over the waters and also over the lunar sphere. Consequently, even the waters of the Deluge did not cause any trouble for Paradise. Enoch, who was already in Paradise even then, did not notice any growth of the waters because of the Deluge. Indeed, the sea is at a higher level than the seashores, as the sight shows. On that account, one must attribute the fact that the sea does not oversteps the limit fixed by God to the divine will. … Let no one misinterpret what I have said at the beginning and believe that I want to teach that the water is not below the heart, by using a language proper to the common people. Is it not actually true that one says that the Antipodes are under our feet? But if you want to speak philosophically, they are under our feet neither more nor less than we are under their feet. But did the Antipodes come perhaps from the remote progenitors? Whereas according to Augustine there are no Antipodes, but usually they are mentioned in theory or as a fiction.[40]

---

[37]Lynn Thorndike, *History of Magic*, op. cit. p. 188.

[38]Ibidem, p. 189.

[39]Ibidem, p. 192.

[40]Alexander of Neckam, *De naturis rerum*, II, XLIX: *Movebitur aliquis super hoc quod dicit propheta, "Dominum firmasse terram super aquas" Ex hoc enim videbitur haberi posse aquas esse inferiores terra, cum tamen Alfraganus dicat, unam esse sphæram aquarum et terræ. Sancti*

3.5 On the Shoulders of Giants. The Thirteenth Century and the Aristotelian Cosmos

"On the shoulders of giants" is the motto which can be taken to characterize the cosmology of the thirteenth century, even if these words appear to have been pronounced in the previous century.[41] In fact, the thirteenth century is that which benefited even more from the translation from the Arabic and the Greek of the works of Ptolemy, Euclid, Aristotle etc. (the giants!). A list of the main translations up to and including the thirteenth century can be found in the second chapter of the work by Alistair Cameron Crombie.[42] The most renowned translators, as we have already mentioned, were Gerard of Cremona (from the Arabic) and William of Moerbeke (from the Greek).[43]

Both Ptolemy's *Almagest* and Aristotle's *De Caelo et Mundo* were the mainstay of the treatments concerning the structure of the cosmos, even if the contradictions between the two proposed models (that of Ptolemy's epicycles and that of Aristotle's spheres) remained for the moment undiscussed. Only later on would the distinction be drawn between the "mathematicians", who made use of the Ptolemaic

igitur expositores referunt illud prophetœ ad cotidianum usum loquendi quo dici solet Parisius fundatam esse super Secanam. Rei tamen veritas est quod paradisus terrestris superior est aquis, cum etiam lunari globo superior sit. Unde et aquœ cataclysmi paradiso nullam intulere molestiam. Enoc, qui in paradiso jam tunc erat collocatus aquarum non sensit diluvii incrementa. Mare vero superius est litoribus, ut visus docet. Unde divinœ jussioni attribuendum est, quod metas positas a Domino non transgreditur mare. ... Quod vero in rubrica dixi nullus perperam intelligat, credens me docere velle aquam non esse sub terra eo loquendi genere quo vulgus uti solet. Nonne enim et antipodes sub pedibus nostris esse dicuntur. Si tamen philosophice loqui volueris, non magis sunt sub pedibus nostris quam nos sub pedibus eorum. Sed numquid de primis parentibus descenderunt antipodes? Secundum Augustinum non sunt antipodes, sed doctrinae causa aut figmenti ita dici solet (our Eng. trans.).

[41]"*Pigmei gigantium humeris impositi plusquam ipsi gigantes vident*", that is, the dwarfs on the shoulders of the giants see more afar then the giants themselves. These words have been attributed to Bernard of Chartres (d. 1130) and, obviously, were referred to his contemporaries who could avail themselves of the ancient philosophers (at that time just translated) for pursuing their research. Even Newton, centuries later, used the same metaphor for saying that he too hold on to the "shoulders" of the ancients. On this subject there is a famous *divertissement* by the sociologist Robert K. Mertons: *On the shoulders of giants* (New York: The Free Press, 1965), reprinted many times.

[42]A. C. Crombie: *Augustine to Galileo:The History of Science, A.D. 400–1650* (William Heinemann, 1952); rpt. as *The History of Science from Augustine to Galileo*, New York: Dover, 1959).

[43]We note that the first translations from Greek to Latin of the works of Aristotle and Plato date back to Boethius and Calcidius and that the Arabs had translated the fundamental Greek works starting from the ninth and tenth centuries. The main European center of translations was in Toledo (Spain) and there, Gerard of Cremona settled himself and became the outstanding figure. On this, see also H. Haskins: *Studies in the History of Medieval Science*, 2nd ed. (Harvard University Press, 1927), Chap. 1. The Flemish William of Moerbecke translated from the Greek the works of Aristotle (also revising the earlier incomplete translations) following the directions of his friend Thomas Aquinas.

model in the astronomical calculations, and the "philosophers", who made reference to the homocentric spheres model which assumed the planets rotating on circular orbits around the Earth. As we know, this question was destined to remain unsolved even in the centuries to come. But, since we must deal with a much more restricted subject, let us continue striving to understand if the old question of the supercelestial waters survives (and how) and inhibits from starting again from Eratosthenes and Ptolemy.

At this point, however, we must make a consideration beforehand. In the above title we have called the thirteenth century as the century of the Aristotelian cosmos but one must say that, even if the diffusion of Aristotle's works of natural philosophy was enthusiastically accepted in the Faculties of Arts, this was not the case within the ecclesiastical hierarchies. Before the synthesis between Aristotelian philosophy and the Catholic theology of Albert the Great and Thomas Aquinas came to be accepted, there was a bitter anti-Aristotle quarrel on the part of the Catholic theologists (mainly at the University of Paris) led by Bonaventura of Bagnoregio, which reached its climax (in 1277) with the condemnation of no fewer than 219 of Aristotle's propositions by the bishop of Paris, Étienne Tempier. One of the main objections, as one can obviously imagine, concerned the eternity of the world assumed by Aristotle. However, among the masters of Arts the natural philosophy of Aristotle took root and, as we know, remained victorious at least four centuries.

Returning now to our task, let us see (as usual, through some examples) how the "new" natural philosophy influenced cosmological theory as well as the physical reality of supercelestial waters.

3.5.1 *Johannes de Sacrobosco (John of Holywood)*

Let us begin with a figure who cannot be defined a philosopher but rather a mathematician or an astronomer, though not an innovator. In fact, Johannes de Sacrobosco was essentially a compiler and it is due to his talent as a compiler that his publishing fortune lasted until the seventeenth century. We cannot say anything about his life: the lack of biographical notes regarding him is almost complete. It is almost certain that he was born in England (he is also named John of Holywood!) and has studied in Paris, where he became a teacher at the Faculty of Arts and died in 1256.

In the twentieth century, at a distance of about forty years one from each other, two important historians of science, among others up to the present, dealt with Sacrobosco: Lynn Thorndike and Olaf Pedersen. Thorndike edited a critical edition of his work *De Sphaera* (about which we shall speak below) and three medieval commentaries of it.[44] Pedersen wrote a long essay fittingly entitled: *In quest of Sacrobosco*.[45] The outcome of all these researches is that all we know of Sacrobosco are the works that have been attributed to him. In addition to *De Sphaera*, three other works are attributed to him with certainty: *Algorismus*, on arithmetic; *De Anni Ratione*, on the reckoning of the calendar; *Tractatus de quadrante*, on the use of a kind of quadrant. This leads us to the conclusion that we are dealing with a teacher who was able to teach courses which we would call today Insitutions of Astronomy and Institutions of Mathematics, obviously within the bounds of what was known in the Middle Ages. All the above considered, why are we so interested in Sacrobosco and in his *De Sphaera*? The reason is that the work of Sacrobosco, with the additions of commentaries, which over the centuries followed after the other, monopolized the university education on astronomy until the beginning of the seventeenth century.

To give an idea of how things have evolved, it is sufficient to compare the length of the original work (forty pages "footnotes included" in Thorndike's edition, net about 24) and the last edition of the commentary by Christopher Clavius, *In Sphaeram Ioannis De Sacro Bosco Commentarius* (1611, about 350 pages in quarto).[46]

Thus, *De Sphaera*, which is assumed to have been composed ca. 1220–1230, became a handbook of cosmography in the European universities. It was divided into four chapters whose contents were exposed in the Proemium:

> The treatise on the sphere we divide into four chapters, telling, first, what a sphere is, what its center is, what the axis of a sphere is, what the pole of the world is, how many spheres there are, and what the shape of the world is. In the second we give information concerning the circles of which this material sphere is composed and that supercelestial one, of which this is the image, is understood to be composed. In the third we talk about the rising and setting of the signs, and the diversity of days and nights which happens to those inhabiting diverse localities, and the division into climes. In the fourth the matter concerns the circles and motions of the planets, and the causes of eclipses.[47]

Sacrobosco fully represents his century from the point of view of the new scientific culture acquired through the translations of the Greek works. For astronomy his master is Ptolemy (Sacrobosco had also read Al-Farghani, ninth century) and for physics Aristotle, of whom he quotes the explanation to solve the

---

[44]L. Thorndike: *The Sphere of Sacrobosco and its Commentators* (University of Chicago Press, 1949).

[45]O. Pedersen: "In Quest of Sacrobosco", *Journal for the History of Astronomy* XVI (1985): 175–220.

[46]Christophorus Clavius: *In Sphaeram Iohannis de Sacrobosco Commentarius* (Mainz, 1611). We note that Christopher Clavius was a very important figure, a teacher at the Collegio Romano for 45 years (starting in 1563) and substantially responsible for the mathematical and astronomical part of the educational programme of Jesuit schools. On Clavius, see also: James M. Lattis: *Between Copernicus and Galileo* (University of Chicago Press, 1994). On the publishing fortune of *De Sphaera*, see D. Boccaletti: "Quando gli astronomi consultavano la Sphaera", *Sapere* LXXVI (2010): 76–81.

[47]L. Thorndike: *The Sphere of Sacrobosco*, op. cit. p. 118; Latin text at p. 76.

problem of the Earth being covered by water but does not quote the model of the homocentric spheres.

Let us look directly at it:

> The machine of the universe is divided into two, the ethereal and the elementary region. The elementary region, existing subject to continual alteration, is divided into four. For there is earth, placed, as it were, as the center in the middle of all, about which is water, about water air, about air fire, which is pure and not turbid there and reaches to the sphere of the moon, as Aristotle says in his book of *Meteorology*. For so God, the glorious and sublime, disposed. And these are called the "four elements" which are in turn by themselves altered, corrupted and regenerated. The elements are also simple bodies which cannot be subdivided into parts of diverse forms and from whose commixture are produced various species of generated things. Three of them, in turn, surround the earth on all sides spherically, except in so far as the dry land stays the sea's tide to protect the life of animate beings. All, too, are mobile except earth, which, as the center of the world, by its weight in every direction equally avoiding the great motion of the extremes, as a round body occupies the middle of the sphere.[48]

As one can see, the sphericity of the Earth and its positioning and immobility in the centre of the universe are firmly maintained. Further on, also proofs of that are supplied.

In addition—and this is a thing of fundamental importance since *De Sphaera* would be a textbook widely used and assumed as an authority—Sacrobosco quotes the measure of the circumference of the Earth given by Eratosthenes.

Let us look at the passage where he substantiates the immobility of the Earth and quotes the measure of its circumference:

> THE EARTH IMMOBILE.—That the earth is held immobile in the midst of all, although it is the heaviest, seems explicable thus. Every heavy thing tends toward the center. Now the center is a point in the middle of the firmament. Therefore, the earth, since it is heaviest, naturally tends toward that point. Also, whatever is moved from the middle toward the circumference ascends. Therefore, if the earth were moved from the middle toward the circumference, it would he ascending, which is impossible. MEASURING THE EARTH'S CIRCUMFERENCE.—The total girth of the earth by the authority of the philosophers Ambrose, Theodosius, and Eratosthenes is defined as comprising 252,000 stades, which is allowing 700 stades for each of the 360 parts of the zodiac.[49]

At this point we feel obliged to insert a remark. Reading *De Sphaera*, one has the impression of being in presence of a "laic" text, unlike those we have dealt with until now. In *De Sphaera* there is no explicit reference to Genesis. The only allusion, which seems to be placed there more for conformism than for necessity, is the short sentence in the second excerpt: "For so God, the glorious and sublime, disposed". In addition, there is no mention of the supercelestial waters.

Below we find the following passage, in which the Creator is mentioned to provide a similitude regarding the motion of the ninth sphere (*primum mobile*):

---

[48]L. Thorndike: *The Sphere of Sacrobosco*, op. cit., p. 119; Latin text at pp. 78–79.

[49]L. Thorndike: *The Sphere of Sacrobosco*, op. cit., p. 122, Latin text at p. 85. Below, in chapter three, he also mentions the fact of the summer solstice at Syene with a quotation from Lucan: "Syene's never varying shadow".

> Be it understood that the "first movement" means the movement of the *primum mobile*, that is, of the ninth sphere or last heaven, which movement is from east through west back to east again, which also is called "rational motion" from resemblance to the rational motion in the microcosm, that is, in man, when thought goes from the Creator through creatures to the Creator and there rests. The second movement is of the firmament and planets contrary to this, from west through east back to west again, which movement is called "irrational" or "sensual" from resemblance to the movement of the microcosm from things corruptible to the Creator and back again to things corruptible.[50]

Considering that Sacrobosco interlards the second and the third chapter with various quotations from Virgil, Ovid and Lucan, one can say that he was more inclined to quote from literature than from the Bible, or perhaps this is due to the atmosphere of the thirteenth century, as we said above.

Coming back to the old problem of the Earth washed by the water (not supercelestial!), a contemporary of Sacrobosco, Michael Scot (1175–1234?), of whom we know very little, wrote a commentary on Sacrobosco's *De Sphaera* in which *inter alia* he faced the problem of the emergence of the Earth from the water. He still agreed with the Cratesian theory of the ocean surrounding the whole Earth and grappled with the Cratesian theory and the Aristotelian model of the concentric elements.[51]

3.5.2 *Robert Grosseteste*

Another commentary (or, better, a rival textbook) to Sacrobosco's *De Sphaera* was written by Robert Grosseteste (1168–1253), who also went back to Cratesian theory. Grosseteste wrote a number of little treatises,[52] among which *De Sphaera*. There, expounding on the Cratesian theory he says:

> Let there be a great circle going round the body of the earth from one pole to another and another great circle going round the body of the earth along the equator. Corresponding to these two circles, two seas circle the earth. And the sea which circles the earth through the poles is called Amphitrides and the other sea is called Ocean. These two seas divide the earth into four parts of which only one is inhabited.[53]

---

[50]L. Thorndike: *The Sphere of Sacrobosco*, op. cit., p. 123, Latin text at p. 86.

[51]See W. G. L. Randles, "Classical Models of World Geography", op. cit., pp. 23–26.

[52]For these, see Ludwig Baur: *Die philosophischen Werke des Robert Grosseteste, Bishops von Lincoln* (Munster, 1912).

[53]L. Baur, op. cit., p. 24: *Intelligatur circulus magnus cingens corpus terrae sub utroque polo, et alius circulus magnus cingens corpus terrae sub aequinoctiali circulo, secundum situm horum duorum circulorum cingunt duo maria totam terram; et illud, quod cingit terram sub polis, amphitrites vocatur, reliquum vero vocatur occeanus. Haec duo maria dividunt terram in quattuor partes, quarum una sola inhabitatur* (Eng. trans. W. G. L. Randles, "Classical Models of World Geography"op. cit. pp. 25–26).

He does not, however, dwell any further on the four parts. With regard to the sphericity of the Earth and of all the stars and planets, he declares that it "is shown both by natural reasons and astronomical experiences", that is, in the case of the Earth, by the observations of the sky by men in different parts of the Earth.

The importance of Grosseteste in the history of science and philosophy does not depend, obviously, only on *De Sphaera*. Even though strictly speaking it is outside our scope, we cannot avoid quoting some biographical notes and indicative excerpts from his works. The biographical notes on the first part of the life of Grosseteste are uncertain and fragmentary. Presently it is believed that he was born in 1168 (formerly, the date believed correct was 1175) in the country of Suffolk. It is not known where he studied the arts, though Oxford seems to be the most probable place. It is also presumed that, afterwards, he was some years in Hereford as well as Cambridge before finally settling in Oxford. It is not sure if, during the closure of the Oxford University from 1209 to 1214, he went to Paris, as other masters of Oxford did. Anyway, it is certain that from 1214 to 1235 he remained at Oxford, first as a master and then later as an ecclesiastic. Around 1229–1230 his name is connected with the institution of a *studium* of Franciscans at Oxford in which he was lecturer of theology. In 1235 he was elected bishop of Lincoln, where he died in 1253. Obviously, we are not interested in his studies of either theology or logic, which he developed by studying and translating Aristotle's works; rather, we want to point out that the natural inquiry had a significant place in the work of Robert Grosseteste.

Even if at present one thinks that the thesis of A. C. Crombie,[54] who had put Grosseteste as a founder of the experimental science, must be reappraised, in any case one must admit that he maintained that the study of nature must be founded on mathematics. In fact, he says:

> The usefulness of considering lines, angles, and figures is very great, since it is impossible to understand natural philosophy without them. They are useful in relation to the universe as a whole and its individual parts. They are useful also in connection with related properties, such as rectilinear and circular motion. Indeed, they are useful in relation to activity and receptivity, whether of matter or sense; and if the latter, whether of the sense of vision, where activity and receptivity are apparent, or of the other senses, in the operation of which something must be added to those things that produce vision.[55]

---

[54]In A. C. Crombie: *Robert Grosseteste and the Origins of Experimental Science, 1100–1700* (Oxford: Clarendon Press, 1953).

[55]*Utilitas considerationis linearum, angulorum et figuraram est maxima, quoniam impossibile est sciri naturalem philosophiam sine illis, Valent autem in loto universo et partibus eius absolute. Valent etiam in proprietatibus relatis, sicut in motu recto et circulari. Valent quidem in actione et passione, et hoc sive sit in materiam sive in sensum; et hoc sive in sensum visus, seeundùm quod oecurrit, sive in alios sensus in quorum actione oportet addere alia super ea, quae faciunt visum* (Latin text from *De lineis, angulis et figuris seu de fractionibus et refiectionibus radiorum* in L. Baur, op. cit., pp. 59–60; trans. David C. L. Lingberg, in *A Source Book in Medieval Science*, op. cit., p. 385).

As one can see, the beginning of this passage calls to mind the famous passage (published four centuries later) of Galileo's *Assayer* (1623). Anyway, despite this statement, whereas Sacrobosco (in the beginning of Chap. I) quotes the definitions of the sphere given by Euclid (322–285 BC) and Theodosius of Bithynia (second–first century BC), Grosseteste does not need to make his *De Sphaera* more "mathematical".

In addition to this, we also want to point out that he also stretched his wings in the field of natural philosophy. In fact, Grosseteste, situated into the context of his time, which provided for various works addressed to optics and the physics of vision, in his treatise *De luce seu de inchoatione formarum* delineates a new cosmogony. Let us look at three indicative excerpts:

> The first corporeal form which some call corporeity is in my opinion light. For light of its very nature diffuses itself in every direction in such a way that a point of light will produce instantaneously a sphere of light of any size whatsoever, unless some opaque object stands in the way. … Thus light, which is the first form created in first matter, multiplied itself by its very nature an infinite number of times on all sides and spread itself out uniformly in every direction. In this way it proceeded in the beginning of time to extend matter which it could not leave behind, by drawing it out along with itself into a mass the size of the material universe. This extension of matter could not be brought about through a finite multiplication of light, because the multiplication of a simple being a finite number of times does not produce a quantity, as Aristotle shows in *De Caelo et Mundo*. However, the multiplication of a simple being an infinite number of times must produce a finite quantity, because a product which is the result of an infinite multiplication exceeds infinitely that through the multiplication of which it is produced. Now one simple being cannot exceed another simple being infinitely, but only a finite quantity infinitely exceeds a simple being. For an infinite quantity exceeds a simple being by infinity times infinity. Therefore, when light, which is in itself simple, is multiplied an infinite number of times, it must extend matter, which is likewise simple, into finite dimensions. … To return therefore to my theme, I say that light through the infinite multiplication of itself equally in all directions extends matter on all sides equally into the form of a sphere and, as a necessary consequence of this extension, the outermost parts of matter are more extended and more rarefied than those within, which are close to the center. And since the outermost parts will be rarefied to the highest degree, the inner parts will have the possibility of further rarefaction. In this way light, by extending first matter into the form of a sphere, and by rarefying its outermost parts to the highest degree, actualized completely in the outermost sphere the potentiality of matter, and left this matter without any potency to further impression. And thus the first body in the outermost part of the sphere, the body which is called the firmament, is perfect, because it has nothing in its composition but first matter and first form. It is therefore the simplest of all bodies respect to the parts that constitute its essence and with respect to its quantity which is the greatest possible in extent. It differs from the genus body only in this respect, that in it the matter is completely actualized through the first form alone. But the genus body, which is in this and in other bodies and has in its essence first matter and first form, abstracts from the complete actualization of matter through the first form and from the diminution of matter through the first form.[56]

[56]L. Baur op. cit. pp. 51, 52, 54: *Formam primam corporalem, quam quidam corporeitatem vocant, lucem esse arbitror. Lux enim* per se *in omnem partem se ipsam diffundit, ita ut a puncto lucis sphaera lucis quamvis magna subito generetur, nisi obsistat umbrosum. … Lux ergo, quae est prima forma in materia prima creata, seipsam per seipsam undique infinities mnltiplicans et in omnem partem aequaliter porrigens, materiam, quam relinquere non potuit, secum distrahens in*

From the conception of the light as original corporeity, active element and form of the matter, he derives the conviction that by means of the laws of optics, in their turn founded on geometry, it is possible to describe and interpret natural phenomena. As one can see, a theory is enunciated which could be defined a metaphysics of light and there has been no lack of those who, always in quest of elements for finding in the works of the ancient philosophers anticipations of the present-day theories, have discerned in the enunciation of Grosseteste the theory of the Big Bang *in nuce*.

3.5.3 Bartholomew the Englishman (Bartholomaeus Anglicus)

Another author who, like Grosseteste, dealt with light (a subject much debated at that time) was Bartholomew the Englishman, in the eight books of his encyclopedic work *De genuinis rerum coelestium, terrestrium et inferorum proprietatibus libri XVIII*, usually referred to as *De proprietatibus rerum*. The work of Bartholomew follows the tradition of Isidore, Bede, and others. Like these, Bartholomew too refers to the creation according to the Bible and to what was said by Augustine, but he also quotes Plato and Aristotle.

Although the work we are talking about was enormously popular for about three centuries (it was translated into Italian, French, English and Spanish during the fourteenth and fifteenth centuries), little is know of the life of its author. It is known

tantam molem, quanta est mundi machina, in principio temporis extendebat. Nec potuit extensio materiae fieri per finitam lucis multiplicationem, quia simplex finities replicatum quantum non generat, sicut ostendit Aristoteles in De caelo et mundo. *Infinities vero multiplicatum necesse est finitum quantum generare, quia productum ex infinita multiplicatione alicuius in infinitum excedit illud, ex cuius multiplicatione producitur. At qui simplex a simplici non exceditur in infinitum, sed solum quantum finitum in infinitum excedit simplex. Quantum enim infinitum infinities infinite excedit simplex. Lux igitur, quae est in se simplex, infinities multiplicata materiam similiter simplicem in dimensiones finitae magnitudinis necesse est extendere. ... Rediens igitur ad sermonem meum dico, quod lux multiplicatione sui infinita in omnem partem aequaliter facta materiam undique aequaliter in formam sphaericam extendit, consequiturque de necessitate huius extensionis partes extremas materiae plus extendi et magis rarefieri, quam partes intimas centro propinquas. Et cum partes extremae fuerint ad summum rarefactae, partes interiores adhuc erunt maioris rarefactionis suscettibiles. Lux ergo praedicto modo materiam primam in formam sphaericam extendens et extremas partes ad summum rarefaciens, in extima sphaera complevit possibilitatem materiae, nec reliquit eam susceptibilem ulterioris impressionis. Et sic perfectum est corpus primum in extremitate sphaerae, quod dicitur firmamentum, nihil habens in sui compositione nisi materiam primam et formam primam. Et ideo est corpus simplicissimum quoad partes constituentes essentiam et maximam quantitatem, non differens a corpore genere nisi per hoc quod in ipso materia est completa per formam primam solum. Corpus vero genus, quod est in hoc et in aliis corporibus, habens in sui essentia materiam primam et formam primam, abstrahit a complemento materiae per formam primam et a diminutione materiae per formam primam* (Eng. trans. Claire Riedl, from: Grosseteste, *On light* (Marquette University Press, 1942), pp. 10–13.

that he was a Franciscan and lived in Oxford, Paris (ca. 1220) and Magdeburg (ca. 1230); it is believed that he wrote *De proprietatibus rerum* around 1240.

On one hand he was not a philosopher, unlike Grosseteste; according to Lynn Thorndike his work was:

> primarily a brief compilation of passages on the natures and properties of things, which are scattered through the works both of the saints and the philosophers, with the intent of making plainer the enigmas which the Holy Scriptures conceal under the symbols and figures of the properties of natural and artificial objects.[57]

On the other hand, Bartholomew himself, both at the beginning and in closing, clearly states that it is an elementary treatise, textbook, or work of reference for benefit of:

> ...young scholars and the general reader who because of the infinite number of book cannot look up the properties of the objects of which Scripture treats, nor are they able to find quickly even a superficial treatment of what they are after.[58]

And it is just a work of this type that can answer our question, which is: What, besides the lofty arguments discussed by the prominent philosophers, was the conventional wisdom at that time in the restricted world of literate people on the problem of the supercelestial waters? Convinced that a work like this will undoubtedly give a faithfully realistic account of the conventional wisdom at the middle of the thirteenth century, we report in its entirety chap. III from Book VIII (where Bartholomew also discusses the universe and celestial bodies):

> On the aqueous or crystalline heaven (*De coelo aqueo sive crystallino*)
>
> The sixth heaven is aqueous, or crystalline, having been formed by the waters placed over the firmament by the divine power; in fact the authority of the Holy Scripture hands down to us that the waters were placed over the heavens and thus transformed and on this account they stay fixed there. However Bede says that those celestial waters, by divine nature, are hanging over the firmament not as tenuous vapor but because of a certain thin solidity. And thus, for moderating the strength of the firmament, or for containing the heat produced by its whirlwind movement. In fact, Bede held that the heaven is not of igneous nature, as the Platonists claimed. Whence Bede says: "the heaven is thin and of igneous nature, round, equidistant from the centre of the earth". And perhaps for this reason Bede thought that waters were necessary there for slowing down that celestial heat, so that the lower world could not be dissipated by such an inflammation. In fact, some say that the coldness of the star of Saturn derives from the natural coldness of those waters placed over the summit of the firmament, because of its closeness to it. They also say that the firmament, cooled by virtue of those waters, cools the orb of Saturn which is closer. But how this can happen rationally is not clearly manifest to those who reason. In fact, since the substance of water by reason of both its qualities, that is, the humidity and the coldness, is almost fully adverse to the celestial substance, it is far from clear to philosophers how among such different bodies it is possible to achieve a certain kind of unity or concordance. However it is written: "He who generates concordance in his highest regions" (Job 12). Hence the moderns think

[57]L. Thorndike, *History of Magic*, op. cit. p. 402.
[58]Ibidem.

in another way, by analysing more deeply, as I think, the interior matter of philosophy. In fact Alexander[59] says that those waters, which are over the heavens, are not placed there as cold, fluid and humid, or rather as solid, frozen and weighty. In fact these properties are also contradictory among themselves and mutually opposing, but those waters are placed over the firmament under the order of the divine wisdom on the strength of their noblest nature, since they are especially close to the celestial nature. And these undoubtedly are the properties of the (celestial) nature: limpidity and transparency, which are principally and substantially in the nature of water which, for this reason, has affinity both with the empyrean heaven and the firmament. Whence God placed the water low on account of the coldness and the humidity and the other conditions necessary to generation and corruption, but made them transparent, since this was necessary to the conservation of all things. Therefore, he says that the heaven is called aqueous and crystalline by reason of mobility and transparency. In fact, it is transparent as crystal, since it receives from the superior heaven, that is from the empyrean, the light or the plenitude of the brightness and diffuses it in the lower things. Therefore, the heaven, almost invisible and hidden to us, is called crystalline not because it is solid like a crystal, but because it is uniformly bright and transparent. Instead, one calls aqueous what moves in the same way as water due to its thinness and mobility, and that motion transmits the movement to the near heaven and this in turn moves what is nearest to itself. And thus, the heaven which moves the lower things is chiefly protective of their motion, as Alexander says.[60]

---

[59]This is Alexander of Hales (1185–1245), he too Franciscan. He was the first Francisian to hold a chair of theology at the University of Paris. His most famous pupil was Bonaventura and his doctrinal references were Augustine and Aristotle.

[60]Bartholomeus Anglicus Francofurti, MDCI—VIII, III, pp. 378–379: *Sextum coelum est aqueum sive crystallinum, quod ex aquis positis super firmamentum divinitatis potentis est formatum, aquas enim esse super coelos collocatas divinae scripturae autoritas nobis tradit, quae ita sunt levigatae & subtiliatae, quod in materiam coelestem sunt conversae, & ideo permanent ibi fixae. Beda tamen docet quod aquae illae coelestes non vaporali tenuitate, sed gracili quadam soliditate, virtute divina super firmamentum sunt suspensae. Et hoc modo ad impetus firmamenti moderationem, vel ad caloris generati ex eius velocissimo motu repressionem. Opinio non fuit Bedae quod coelum igneae sit naturae, sicut Platonici posuerunt. Unde dicit Beda. Coelum est subtilis & igneae naturae, rotundum a centro terrae, aequalibus spaciis collocatum. Et ideo forte Bedae videbatur quod ideo fuit necesse ibi esse aquas, ut calor ille coelestis ad temperantiam duceretur, quod ex tali inflammatione mundus inferior dispendium non pateretur, ex frigiditate enim naturali illarum aquarum super firmamenti verticem positarum, dicunt aliqui stellam Saturni esse frigidam propter illam, quam habet ratione situs cum firmamento vicinitatem. Dicunt etiam quod firmamentum per virtutem illarum aquarum infrigidatum infrigidat orbem Saturni sibi magis proximum & vicinum. Sed qualiter istud posset fieri retionabiliter, non est perspicue ratione retentibus manifestum. Nam cum aquosa substantia ratione utriusque qualitatis suae scilicet humiditatis & frigititatis, coelesti substantiae penitus sit contraria, non est bene liquidun philosophantibus, qualiter inter corpora tam disparia possit unitas aut concordia aliqualiter conveniri. Et tamen scripum est Job 12. Qui facit concordiam in sublimibus suis. Ideo alio modo sentiunt & opinantur moderni, qui interiora philosophiae spectamina profundius, ut arbitrer, sunt scrutati. Dicit enim Alexander, quod aquae illae, quae super coelo sunt, non sunti bi positae ut frigide & fluxibiles & humidae, vel etiam sicut solidae, congelatae, & ponderosae. Istae enim proprietates sunt etiam inter se habentes contrarietatem, & sibi mutuo repugnantes, sed potius per ordinationem divinae sapientiae aquae illae super firmamentum sunt sub nobilissima naturae suae conditione divinitus collocatae, prout naturae coelesti maxime sunt propinquae. Et haec quidem est naturae proprietas perspiquitatis & transparentiae, quae principaliter & substantialiter invenitur in natura aquae, ratione cuius habet convenientiam & cum coelo empyreo & etiam cum firmamento. Et ideo posuit Dominus aquas inferius sub ratione frigidi & humidi, cum aliis conditionibus necessariis ad*

Therefore, concerning the supercelestial waters, the authority of Scripture causes Bartholomew to accept the existence of an aqueous and crystalline heaven, but he rejects Bede's view that these waters are cold and congealed. It is not because they are congealed but because they are transparent that this heaven is called crystalline. In other words, concludes Thorndike, the "supercelestial waters" are not really waters. And we too can conclude that the Scholastic debate on the supercelestial waters faded by dematerializing the waters.

3.5.4 Albert the Great and Roger Bacon

In the second half of the thirteenth century, we meet with two thinkers undoubtedly interested in the subject we are dealing with: Albert the Great (Albertus Magnus, 1193–1280) and Roger Bacon (1219–1292), the first a Dominican and the second a Franciscan. Both were interested in the natural sciences and enjoyed the fame of being innovators in this field. This is perhaps unmerited, since in their assertions they based themselves specially on the authority, but it is nevertheless especially true of Bacon, since from the nineteenth century he has been considered a pioneer of the experimental method. In his works he also maintained the importance of mathematics in the natural sciences, but in his work *Opus tertium*, while he says he wants to apply the principles of the geometric science to the things of this world and to show how in the celestial things one finds these rules of behaviour, he ends up "proving" his theses through Aristotle and Pliny.

Bacon expounds an odd theory regarding the light of the Moon: It is not reflected light, it comes precisely from the Moon and this can be proven on the strength of the identity of the angles of incidence and reflection. In any case, it was drawn into the body of the Moon, by virtue of the Sun, by the power of matter.

Both Bacon and Albert expounded their ideas regarding the inhabitability of the oecumene. Let us begin with Albert (in his work *De natura loci*). Since he is fully aware of Aristotle's theory of the elements, he knows that the sphere of water is greater than the sphere of the earth, and that this makes it so that the Earth is not inhabitable everywhere. Besides, according to many philosophers, since the northern hemisphere is not covered by the water, the southern one instead must be covered and thus uninhabitable. Albert denies this conclusion by quoting the results

---

generationem & corruptionem, sed easdem posuit in ratione perspicui prout fuit necessarium ad universitatis consevationem. Et ideo dicit coelom esse dictum aqueum & crystallinum ratione mobilitatis & perspicuitatis. Est enim perspicuum ad modum crystalli, a superiori coelo scilicet ab empyreo, lucem vel fumositais plenitudinem recipiens, receprum ad inferiora diffundens. Et ideo dicitur coelom quasi nobis invisibile & occultum, crystallinum, non quia durum sicut crystallus. Sed quia uniformiter est luminosum & et perspicuum. Aqueum autem dicitur, quemadmodum aqua ex sua subtilitate & mobilitate movetur, & illud motum movet coelom proximum, & illud ulterius movet quod sibi est magis propinquum. Et ideo illud coelom quod movet inferiora, inferiorum mobilium praecipue est conservativum, ut dicit Alexander (our Eng. trans.).

of the astronomers who assert the identity of the two hemispheres. He asserts that the torrid zone is difficult, but not impossible, to cross, and that the southern hemisphere is not only inhabitable, but is inhabited: it is divided up just as ours is, with regions that are habitable or uninhabitable depending on the local temperature.

Analogous conclusions are reached by Bacon (also in his other work *Opus maius*), although his arguments are different.[61] After an initial explanation through the "multiplication of the beams", then he ends up with the help of Aristotle and Pliny.

If from a general philosophical point of view Albert the Great and Roger Bacon mark an important stage in the evolution of Western thought, this is not the case with regard to cosmography (for instance, they did not even realize the contradiction between the models of Aristotle and Ptolemy) and the nature of the Earth. Bacon, following Aristotle, also deals with the problem of the unicity and finiteness of the world. Substantially, he presents nothing new. This judgement is, obviously, strictly limited to the subject Earth. It is true that the thirteenth century is an "Aristotelian" century, but this century also witnessed the translation of the *Almagest* and it seems that no effort was made to improve the understanding of the differences between the two conceptions of the universe.

The most debated subject at that time seems to have been the one we have quoted above when speaking of Albert the Great: the sphere of water is greater than that of the Earth. But the impression one draws from those discussions is that the subject is dealt with as a literary topos: we shall see in the next section the exploit of Dante Alighieri.

3.5.5 Dante and the *Questio de Aqua et Terra*

As we have already recounted in the course of this chapter, the development of Christian philosophical thought in the Middle Ages took place first in the monasteries, then in the cathedral schools and at last in the universities. In this last phase, it was no longer only ecclesiastics but also laymen who took part in the intellectual debates. In this age reappears the figure of the intellectual protected by a powerful person (for instance a lord of an important family), or even directly in his service with various tasks (embassages, etc.).

Dante Alighieri (1265–1321) was one of these. He is not a professor in one of the most famous universities of that time (Oxford, Paris, Bologna, etc.) nor did he belong to any religious order. Today he would be defined an "independent scholar", and as such he participated in the discussions of the "scientific" problems of his time.

During a stay in the city of Mantua (perhaps at the court of the Bonacolsi, allied to Cangrande della Scala), Dante participated in a scientific debate on the

[61]On this, see W. L. Randles, "Classical Models of World Geography", op. cit., pp. 27–28.

cosmologic problem of the reciprocal ratios of the spheres of water and earth, which, as we have seen, was much discussed in medieval culture and in the late thirteenth century aroused particular interest in consequence of the recent translations of Aristotle's works.

In fact, once the Aristotelian cosmological hypothesis, which considered the four spheres of the elements being arranged homocentrically (from the heaviest to the lightest) under the heaven of the Moon, was accepted as a general thesis, it remained to investigate and to solve, in the light of effectual truth and empirical observations, the problem connected to the emergence of the Earth in the so-called "inhabited quarter" and, at the level of scientific hypothesis, that of the ratio between the two spheres, aqueous and terrestrial.

The debate took place in Mantua, a culturally animated town also aware of certain "philosophical" problems; it was, in fact, precisely in Mantua that the work of Bartolomew the Englishman (*De rerum proprietatibus*) had been translated (into Mantuan vernacular) between 1299 and 1309. Dante summarized the result of the debate, or better the solution proposed by him to the questions that had been posed, in a lecture held on 20 January 1320 in the chapel of St. Helen, a little church near the Cathedral of Mantua, and reworked in the form of a letter (first printed in 1508, and for a long time considered spurious).

Here it is how he introduces his speech:

> Dante Alighieri of Florence, least among real philosophers, to each and every reader of these words, greeting in His name, who is the source and beacon of truth. Let it be known to you all that, whilst I was in Mantua, a certain Question arose, which, often argued according to appearance rather than to truth, remained undetermined. Wherefore, since from boyhood I have ever been nurtured in love of truth, I could not bear to leave the Question I have spoken of undiscussed: rather I wished to demonstrate the truth concerning it, and likewise, hating untruth as well as loving truth, to refute contrary arguments. And lest the spleen of many, who, when the objects of their envy are absent, are wont to fabricate lies, should behind my back transform well-spoken words, I further wished in these pages traced by my own fingers to set down the conclusion I had reached and to sketch out, with my pen, the form of the whole controversy. The Question then concerned the location and the shape, or form, of two of the elements, Water namely and Earth; and here I call "form" that which the Philosopher in the Categories ranks as the fourth species of Quality. And the Question was confined to this—as the very foundation of the truth to be investigated—that the inquiry should be: whether water in its own sphere, that is in its natural cir-cumference, is anywhere higher than the earth which emerges from the waters and which we commonly call the "habitable quarter:" and this was argued to be so for many reasons, of which, discarding some for their shallowness, I have retained five which appeared to have some weight.[62]

[62]Dante Alighieri, *Quaestio De Aqua et Terra*: *Universis et singulis praesentes litteras inspecturis, Dantes Aligherius de Florentia, inter vere philosophantes minimus in Eo salutem, qui est principium veritatis et lumen. Manifestum sit omnibus vobis quod, existente me Mantuae, quaestio quaedam exorta est, quae dilatata multoties ad apparentiam magis quam ad veritatem, indeterminata restabat. Unde quum in amore veritatis a pueritia mea continue sim nutritus, non sustinui quaestionem praefatam linquere indiscussam: sed placuit de ipsa veram estendere, nec non argomenta facta contra dissolvere, tum veritatis amore tum etiam odio falsitatis. Et ne livor multorum qui absentibus viris invidiosis mendacia confingere solent, post tergum benedicta*

Dante proceeds to list and explain the five points in which he has schematized the reasoning of his opponents. Then, having put in Chap. VIII a fundamental empiric assessment (the course of the rivers clearly tends downwards), he continues by maintaining his intention of substantiating it with strictly and rigorously rational arguments. In fact, in Chap. IX he says:

> This will be the order. First, we shall demonstrate that it is impossible for water in any part of its circumference to be higher than the emergent or uncovered land. Secondly, we will show that this emerging land is everywhere higher than the whole surface of the sea. Thirdly, we shall argue against our conclusion and refute the objections. Fourthly, we shall show the final and efficient cause of this elevation or emerging of the land. Fifthly, we shall refute the arguments above set down.[63]

The first four parts of the discussion occupy the Chaps. X–XXII. The fifth part (Chaps. XXIII–XXIV) is finally addressed to a summary examination and a sharp and accurate criticism of the five "proofs" presented by the opponents, sharply contrasted through a rapid and precise summing up discourse. We report the closing (Chap. XXIV):

> This philosophical inquiry was held, beneath the rule of the unconquered lord, the Lord Can Grande della Scala, Vicar of the Holy Roman Empire, by me, Dante Alighieri, least among philosophers, in the famous town of Verona, in the Chapel of the glorious Helen, before the entire Veronese clergy, save a few, who, burning with excess of love do not admit the inquiries of others, and, by virtue of their humility, poor in the Holy Spirit, lest they should appear to testify to the worth of others, avoid attending their discourses. And this was done in the year 1320 after the birth of our Lord Jesus Christ, on Sunday, which our predestined Saviour bade us keep holy because of his glorious nativity and his marvellous resurrection;

---

transmutent, placuit insuper in hac cedula meis digitis exarata, quod determinatum fuit a-me relinquere, et formam totius disputationis calamo designare. Quaestio igitur fuit de situ et figura, sive forma duorum elementorum, Aquae videlicet et Terrae ; et voco hic formata illam, quam Philosophus ponit in quarta specie qualitatis in Praedicamentis. Et restricta fuit quaestio ad hoc, *tamquam ad principium investigandae veritatis, ut quaereretur: Utrum aqua in spherae sua, hoc est in sua naturali circumferentia, in aliqua parte esset altior terra quae emergit ab aqnis, et quam communiter quartam habitabilem appellamus; et arguebatur quod sic multis rationibus, quarum (quibusdam amissis propter earum levitatem) quinque retinui, quae aliquam efficaciam habere videbantur* (Eng. trans. Alain Campbell White: *A translation of the* Quaestio De Aqua et Terra, *with a discussion of its authenticity*. The Latham Prize essay, 1902, pp. 2–6).

[63]Dante Alighieri, *Quaestio De Aqua et Terra*: IX. *Hic erit ordo. Primo demonstrabitur impossibile, aquam in aliqua parte suae circumferentiae altiorem esse hac terra emergente sive detecta. Secando demonstrabitur, terram hanc emergentem esse ubique altiorem totali superficie maris. Tertio instabitur contra demonstrata, et solvetur instantia. Quarto ostendetur causa finalis et efficiens huius elevationis sive emergentiae terrae. Quinto solvetur ad argumenta superius praenotata* (Eng. trans. Alain Campbell White, op. cit, pp. 14–15).

which day was the seventh after the ides of January and the thirteenth before the calends of February.[64]

Alas, we could not still further weigh our discourse with other excerpts, but we point out that an inspection of this little work shows the absence of quotations from Genesis or from the Fathers of the Church; the only authority quoted many times is Aristotle. The hurdle of the Bible has been overcome.

3.6 The Fourteenth Century and the New Interpretations of Aristotles's Theory of the Natural World

As we have seen, in the course of time the problem of rationally explaining the narration of Genesis regarding the presence of water over the firmament has given way to the necessity of substantiating (always rationally) the Aristotelian model of the four elements. In the aforementioned case of Dante, the problem even concerned only two of them: *quaestio de duobus elementis aquae at terrae*. One can perhaps conclude that the problem faded away with the demise of the Scholastics. Other philosophical movements were born bringing with themselves new schools of thought. In the fourteenth century, two important schools flourished in Europe, in the universities of Oxford and Paris, respectively. Both these schools made important contributions to the progress of mathematics and natural sciences, particularly of mechanics. As usual, we shall only deal with the novelties concerning the Earth. As far as kinematics and dynamics (in general and the impetus theory) are concerned, we refer the reader to our *Galileo and the Equations of Motion*.[65]

We owe to the most important of the philosophers of Merton College of Oxford, Thomas Bradwardine (1290–1349), the calculations regarding the proportion between the volumes of the four elements. What was the problem? In the Ptolemy's *Almagest* the hypothesis had been presented that the distance between the Earth and the Moon was approximately 33.3 times the radius of the Earth. This evaluation had also been reported by the Persian astronomer Alfraganus (al-Farghani, ninth century) in his treatise *Elements of Astronomy*, translated into Latin by John of Seville in 1137 under the title of Differentia Scientie Astrorum and read by Bradwardine.

---

[64]Dante Alighieri, *Quaestio De Aqua et Terra*: XXIV. *Determinata est haec philosophia dominante invicto Domino, domino Kane Grandi de Scala pro Imperio sacrosancto Romano, per me Dantem Aligherium, philosophorum minimum, in inclyta urbe Verona, in sacello Helenae gloriosae, coram universo clero Veronensi, praeter quosdam qui, nimia cantate ardentes, aliorum rogamina non admittunt, et per humilitatis virtutem Spiritus Sancti pauperes, ne aliorum excellentiam probare videantur, sermonibus eorum interesse refugiunt.—Et hoc factum est in anno a nativitate Domini nostri lesu Christi millesimo trecentesimo vigesimo, in die Solis, quem praefatus noster Salvator per gloriosam suam nativitatem, ac per admirabilem suam resurrectionem nobis innuit venerandum; qui quidem dies fuit septimus a lanuariis idibus, et decimus tertius ante Kalendas Februarias* (Eng. trans. Alain Campbell White, op. cit, pp. 56–58).

[65]D. Boccaletti: *Galileo and the Equations of Motion* (Springer, 2016), Chap. 2.

However, at the time of Bradwardine, it was also known that some commentators of Aristotle's *On the Heavens* held that the ratio between the volume of an element and that of the preceding element was 10:1. Bradwardine, suspecting that the two hypotheses could be contradictory, formulated a calculation for determining the ratio of proportionality (which he continued to consider constant) between the four volumes. Simplifying, he uses 33 in place of 33.3 and arrives at the obvious (for us!) result that the volume contained within the surface of the orb of the Moon must be $33^3 = 35{,}937$ times the volume of the Earth.

Always assuming a continuous proportionality between the volumes of the four elements, Bradwardine succeded in evaluating (approximately) the constant ratio between them. He required that the sum of the volumes of the four elements contained within the orb of the Moon be equal to 35,937, considering 1 the volume of the Earth: the result, obtained by trial and errors, gave a figure between 32 and 33.

The figure obtained was obviously very different from that proposed by the commentators of Aristotle! The calculation is contained in the third part of the fourth chapter of his *Tractatus de Proportionibus* (1328) where he says that he will uncover the secret of the ratio between the elements even if this is not so relevant to the subject of the treatise.[66]

According to historians, from this he won a reputation as a great mathematician among his contemporaries but things did not change. Afterwards, in fact, the astronomers continued to use the famous ratio 33.3 between the radii of the Earth and of the orb of the Moon, and the Aristotelian philosophers continued to trust in the ratio 10:1 between the volumes.

3.6.1 Jean Buridan

Let us pass now to the University of Paris, where, in about the same years of Bradwardine at Oxford, the most distinguished and influential teacher of natural philosophy was Jean Buridan (ca. 1295–ca. 1358).[67] As Moody says,

> Buridan made significant and original contributions to logic and physics, but as a philosopher of science he was historically important in two respects. First, he vindicated natural philosophy as a respectable study in its own right. Second, he defined the objectives and methodology of scientific enterprise in a manner that warranted its autonomy with respect to dogmatic theology and metaphysics.[68]

[66]See H. Lamar Crosby: *Thomas Bradwardine, Tractatus de Proportionibus. Its significance for the Development of Mathematical Physics* (University of Wisconsin Press, 1961), pp. 133–141.

[67]See E. A. Moody: "Buridan, Jean" in *Dictionary of Scientific Biography*, vol II (1970), pp. 603–608.

[68]E. A. Moody, "Buridan, Jean", op. cit., p. 604.

In his teaching, Buridan dealt with logic and natural sciences, first and foremost through commentaries on the works of Aristotle, from whom he sometimes diverged, for instance with the impetus theory, which became famous later on.

The subject we are interested in was dealt with by Buridan essentially in the work *Quaestiones de Caelo et Mundo*,[69] i.e., in a commentary on Aristotle's *On the Heavens*. From it, we consider first the *Quaestio Septima* of the second book bearing the indicative title "*Utrum tota terra sit habitabilis*", where Buridan argues against the Cratesian theory.

After having said "*Ista quaestio videtur mihi valde difficilis*", the habitability of the Earth is first discussed on the basis of the division into the five climatic zones: according to traditional opinion the tropical and the polar zone were not habitable because of their lack of a temperate climate; according to Avicenna, instead, the tropical zones would enjoy a mild climate. He then passes to a discussion of the habitability of the Earth in proportion with the distribution of the waters. We quote the relevant passage:

> Now, it remains to speak of uninhabitation caused by waters. There are three great opinions on this. Some assume that only one quarter is habitable, while others assume that all quarters are habitable. And this [latter] opinion will be discussed first. They say that both earth and water are concentric to the world, so that the center of the world is the center of both of them. However, they say that in any quarter of the earth there are many regions not covered by waters because of many protrusions of earth and elevations of mountains projecting above the waters. And they [also] say that many other parts of the earth are covered by waters because of their depressions, such as valleys between the aforementioned elevations. They say that is so in any quarter of the earth. The sign of this is that from one very large uncovered region, we cross a great and long sea and come to [yet] another very large uncovered region. It is probable that this would be the case as one went round the whole earth. The first is that all the seas which can be crossed and all the habitable lands which can be found are contained in this quarter of the earth which we inhabit. And some have tried to cross the sea into other quarters but they could never reach any habitable land. It is said, therefore, that Hercules fixed pillars at the limits of this quarter as a sign that beyond them there was no habitable land and no passable sea. The other doubt, which has been previously stated, is more difficult, because if the world is eternal, this opinion cannot explain how the earth's elevations can be saved through eternity, since from these elevations many parts of earth always flow with the rivers into the depths of the sea. Through an infinite time the depths of these seas ought to be filled and the elevations of the earth ought to be consumed, which is not a convenient thing to tell those who wish to maintain the eternity of the world in a condition as favorable to animals and plants as at the present time.[70]

[69]See: Johannis Buridan: *Quaestiones super Libris quattuor de Caelo et Mundo*, ed. E. A. Moody, (Cambridge MA: The Medieval Academy of America, 1942).

[70]*Quaestiones super ...*, op. cit. pp. 157–158: *Nunc restat dicere de inhabitatione propter aquas. Et sunt de hoc tres magnae opiniones. Aliquid ponunt unam solam quartam vel quasi habitabilem, et alii ponunt omnes quartas terrae habere habitationem. Et de ista opinione erit primo dicendum. Isti ergo dicunt tam terram quam aquam esse concentricas mundo, ita quod centrum mundi sit centrum earum ambarum tamen dicunt in qualibat quarta terrae esse multas plagas discoopertas aquis, propter multas terrae gibbositates et quasi montium elevationes eminentes super aquas. Et dicunt multas alias partes terrae esse coopertas aquis propter earum depressions, ad modum vallium inter praedictas elevationes, Et hoc dicunt ita esse in qualibet quarta terrae; cuius signum*

As a comment on this excerpt, we subscribe to the remark by William Randles:

> Should we be surprised to see this university scholar rely on the myth of Hercules and his pillars to refute the Cratesian theory, when navigation beyond the straits of Gibraltar had for long been regularly practised by Mediterranean peoples and a Portuguese expedition led by the Genoese Lanzarotto Malocello had in 1336 reached the Canary islands probably shortly before Buridan wrote? As will be continually evident further on, the information furnished by theoretically oriented scholars and practical mariners was frequently out of phase, and lack of communication between them, except in rare cases, was a constant feature in our context.[71]

We can add as a further consideration that, in spite of the progress made, the natural sciences now and then deal with a nature more imagined than known. One can also add, quoting the ironic title by John Murdoch,[72] that the natural philosophers of the Middle Ages practiced a natural philosophy without nature.

Buridan then continues to discuss the problem of the safety of living beings and plants by presenting the possible solutions (divine intervention included) and the opposing opinions. Finally he proposes:

> … there is a third opinion, which seems probable to me and by means of which all appearances could be perpetually saved. Let both the earth and water be assumed concentric to the world so that the whole earth naturally collects around the center of the world and all water naturally flows down toward a lower place with respect to the center of the world. But there is much water in the bowels of the earth and through evaporation there is also much water mixed together with air. Hence it is not necessary that there be so much water in the sea that it should exceed the elevations of the earth. But now we must ask how these elevations will be saved eternally. If, according to Aristotle, the world is assumed eternal, it could be replied that from eternity the world has been ordained for the well-being of animals and plants because one part of the earth, nearly a fourth, has not been covered by waters and rises above the waters; and it always remains, and will remain, naturally

---

est, quod de una plaga valde magna discooperta nos pertransimus valde magnum et longum mare et venimus ad aliam plagam discopertam valde magnam, et verisimile est quod ita esset circumeundo terram totam. Sed contra istam opinionem sunt duae magnae dubitationes. Prima est, quia omnia maria quae ab aliquibus poterunt transiri, et omnes terrae habitabiles quae poterunt inveniri, continentur in ista quarta terrae quam habitamus. Et aliqui laboraverunt in mari ad permeandum mare in aliis quartis, et nunquam potuerunt pervenire ad aliquam terram habitabilem; et ideo dicitur quod Hercules in finibus huius quartas infixit columnas, in signum quod ultra eas non erat terra habitabilis nec mare permeabile. Alia dubitatio difficilior est, quae dicta fuerit primus, quia haec opinio non potest salvare, si mundus fuerit aeternus, quo modo istae elevationes terrae possunt salvari ab aeterno, cum semper ex eis fluant multae partes terrae cum fluviis ad fundum maris. Iam enim ab infinito tempore deberent tales fundi marium esse repleti, et deberent elevationes terrarum esse consumptae; quod non est conveniens dicere volentibus tenere perpetuitatem mundi in statu prospero animalibus et plantis sicut nunc est (Eng. trans. Edward Grant, *Source Book in Medieval Science*, op. cit., p. 622).

[71]W. G. L. Randles, "Classical Models of World Geography", op. cit., p. 32.

[72]John E. Murdoch, "The Analytic Character of Late Medieval Learning: Natural Philosophy without Nature", in: Approaches to Nature in the Middle Ages: Papers of the Tenth Annual Conference of the Center for Medieval and Early Renaissance Studies, L. D. Roberts, ed. (Binghamton NY: Center for Medieval and Renaissance Studies and Text, SUNY Binghamton, 1982), pp. 171–213.

> uncovered despite the concentricity and even though we might confine [or restrict] the mountains. And there is a conception *(imaginatio)* that in the uncovered part the earth is altered by air and the sun's heat, and much air is mixed with it, so that this earth becomes rarer and lighter and has many pores filled with air or subtle bodies. However, the part of the earth covered with waters is not altered by air and sun and therefore remains denser and heavier. And therefore, if the earth were divided through the middle [or center] of its magnitude, one part would be much heavier than another, but that part which is uncovered would be much lighter. It seems, then, that there is one center of magnitude of the earth and another center of gravity. For the center of gravity is where the heaviness is just as much on one side as on the other, but as was said, this is not in the middle of the magnitude. Furthermore, since by its heaviness the earth tends to the middle of the world, its center of gravity is the middle of the world and not its center of magnitude. It is because of this that the earth is raised above the water on one side and is wholly under water on the other side.[73]

This "contamination" of the Aristotelian theory of the four elements with the Archimedean concept of centre of gravity had a notable success (successively also being supported by Albert of Saxony, a contemporary and colleague of Buridan at the University of Paris), which lasted until the early sixteenth century, when the undertakings of the Portuguese sailors demonstrated the great extent of the land in the southern hemisphere.

Let us pass now to the *Quaestio XXII—Utrum terra semper quiescat in medio mundi* (whether the Earth be always still in the midst of the world). Here, Buridan finally tackles a new problem (new, obviously, for the Middle Ages since, as we know, already in the III century BC Aristarchus of Samos had presented his heliocentric theory, in which the Earth performs a rotation on its axis and a revolution around the Sun).

---

[73]*Quaestiones super ...*, op. cit. p. 159: *...est tertia opinio, quae videtur mihi probabilis, et per quam perpetuo salvarentur omnia apparentia, ponendo quod tam terra quam aqua sunt concentricae mundo, ita quod tota terra est innata congregari circa centrum mondi, et etiam omnia aqua est innata fluere ad locum decliviorem respectu centri mundi. Sed multa aqua est in visceribus terrae, et multa etiam est commixta aeri per evaporationes; ideo non oportet tantam aquam esse in mari quod excedat elevationes terrae. Sed tunc quaeritur, quo modo aeternaliter salvabuntur illas elevationes terras. Respondetur, si secundum Aristotelem poneretur mundus aeternus, quod ab aeterno ad salutem animalium et plantarum mundus est ordinatus quod una pars terrae, quasi una quarta, est discooperta aquis et eminens super aquas; et semper manet, et manebit etiam naturaliter discooperta, non obstante concentricitate et licet etiam circumscriberemus montes. Et est talis imaginatio, quod terra in parte discooperta alteratur ab aere et a calore solis, et commiscetur sibi multus aer, et sic fit illa terra rarior et levior et habens multos poros repletos aere vel corporibus aubtilibus; pars autem terrae cooperta aquis non sic alteratur ab aere et sole, ideo remanet densior et gravior. Et ideo si divideretur terra per medium suae magnitudinis, una pars esset valde gravior quam alia, illa autem pars in qua terra esset discooperta esset multo levior. Et sic apparet quod aliud est centrum magnitudinis terrae, et aliud est centrum gravitatis eius; nam centrum gravitatis est ubi tanta est gravitas ex una parte sicut ex altera, et hoc non est in medio magnitudinis, ut dictum est. Modo ultra, quia terra per suam gravitatem tendit ad medium mundi, ideo centrum gravitatis terrae est medium mundi, et non centrum suae magnitudinis; propter quod terra ex una parte est elevata supra aquam, et ex alia parte est tota sub aqua* (Eng. trans. Edward Grant, in *Source book in Medieval Science*, op. cit., pp. 622–623).

After having explained the reason in support of the terrestrial motion and said also in this case that the question is difficult, Buridan says:

> It should be known that many people have held as probable that it is not contradictory to appearances for the earth to be moved circularly in the aforesaid manner, and that on any given natural day it makes a complete rotation from west to east by returning again to the west—that is, if some part of the earth were designated [as the part to observe]. Then it is necessary to posit that the stellar sphere would be at rest, and then night and day would take place through such a motion of the earth, so that that motion of the earth would be a diurnal motion *(motus diurnus)*. The following is an example of this [kind of thing]: If anyone is moved in a ship and he imagines that he is at rest, then, should he see another ship which is truly at rest, it will appear to him that the other ship is moved. This is so because his eye would be completely in the same relationship to the other ship regardless of whether his own ship is at rest and the other moved, or the contrary situation prevailed. And so we also posit that the sphere of the sun is everywhere at rest and the earth in carrying us would be rotated. Since, however, we imagine that we are at rest, just as the man located on the ship which is moving swiftly does not perceive his own motion nor the motion of the ship, then it is certain that the sun would appear to us to rise and then to set, just as it does when it is moved and we are at rest.[74]

One can remark that, whereas the opinions of the Fathers of the Church are always quoted with the name of the author, in this case Buridan confines himself to say "*multi tenuerunt*" without further specifying. One can perhaps think that, by referring to the heliocentric hypothesis without naming an author, he wants to defuse the hypothesis itself even if he has straight that, since it is a question of relative motion, the appearance is the same.

He will also say later on "We should know likewise that those persons wishing to sustain this opinion, perhaps for reason of disputation, posit for it certain persuasions",[75] where "*gratia disputationis*" counterbalances the "*persuasiones*", i.e., the heliocentric hypothesis is presented by some people as a provocation. As the saying goes, you can never be too careful.

As far as we know, Buridan was the first, in the Middle Ages, to present an example of relative motion observed from the two relative frames of reference.

---

[74]*Quaestiones super…*, op. cit., p. 227: *Sciendum est ergo quod multi tenuerunt tanquam probabile, quod non contradicit apparentibus terram moveri circulariter modo praedicto, et ipsam quolibet die naturali perficere unam circulationem de occidente in orientem revertendo iterum ad occidentem—scilicet si aliqua pars terrae signaretur. Et tunc oportet ponere quod sphaera stellata quiesceret, et tunc per talem motum terrae fierent nobis nox et dies ita quod ille motus terrae esset motus diurnus. Et potentis de hoc accipere exemplum; quia si aliquis movetur in navi et imaginetur se quiescere, et videat aliam navem quae secundum veritatem quiescit, apparebit sibi quod illa alia navis moveatur; quia omnino taliter se habebit oculus ad illam aliam navem, si propria navis quiescat et alis moveatur, sicut se haberet si fieret e contrario. Et ita etiam ponamus quod sphaera solis omnino quiescat, et terra portando nos circumgiretur cum tamen imaginemur nos quiescere, sicut homo existens in navi velociter mota non percipit motum suum nec motum navis, tunc certum est quod ita sol nobis oriretur et postea nobis occideret sicut modo facit quando ipse movetur et nos quiescimus* (Eng. trans. Marshall Clagett, in *Source Book in Medieval Science*, op. cit., p. 501).

[75]*Quaestiones super…*, op. cit., p. 227: *Debemus etiam scire quod volentes istam opinionem forte gratis disputationis sustinere, ponunt ad eam quasdam persuasiones.*

The recourse to the example of the motion of a ship was followed by others in the course of time: by Nicole Oresme[76] in 1377 and later on by Nicolas of Cusa[77] and René Descartes,[78] and finally by Galileo, with the famous excerpt of the second day of his *Dialogue.*[79] And Galileo, as we know, was less cautious than Buridan.

To tell the truth, Oresme (1323–1382) went even farther than Buridan in suggesting arguments in support of the rotation of the Earth, but in the end, having reached the conclusion that both the solutions (rotation of the heaven or rotation of the Earth) had equal possibilities, he opted for the first, yielding to the Bible. Oresme, author of various works of commentary on Aristotle (just above we mentioned his commentary on *On the Heavens*, written in French at the instance of Charles V King of France) can be considered, as regards natural philosophy, together with Buridan, the point of arrival of medieval philosophical research.

Natural philosophy does not represent the leading part of philosophical research in the ten centuries which go under the name of Middle Ages,[80] but for our purposes, one can affirm that, in addition to recovering (with difficulty!) the notion of the sphericity of the Earth, it has introduced a new question into the intellectual debate: the conflict or agreement between faith and reason. From now on, research in physics and astronomy will have to come to terms with what is maintained in the Bible.

As a result of the survey carried out in this chapter, we can say that the last centuries of the Middle Ages have returned to us a spherical Earth immobile in the centre of the universe in the context of a "mixed" cosmological theory formed by Aristotle's model of the homocentric spheres and by Ptolemy's mathematical model, destined to live parallel lifes. As we have already anticipated, no solution of the problem will be reached, not least because the "users" of the two models will be different: the "philosophers" will use the Aristotelian model, the "mathematicians" (i.e., the astronomers) will use the Ptolemaic model.

In the legacy of the Middle Ages, as we have seen, also included is the question of the distribution of the waters over the Earth (connected to the distribution of the four elements according to Aristotle) and a timid and problematic appearance of the terrestrial rotation. Given the more "conceptual" than "practical" nature of the considerations which concerned the Earth during the Middle Ages in the West, the question of the size of the Earth was never faced directly, nor were methods of measurement proposed. Instead, in the East during the same period some projects of

---

[76]Nicole Oresme: *Le livre du ciel et du monde*, eds. A. D. Menut and A. J. Denomy (University of Wisconsin Press, 1968), p. 523.

[77]Nicholas of Cusa: *De docta ignorantia*, II, 12.

[78]René Descartes: *Principia Philosophiae*, II, XV, 1644.

[79]Galileo Galilei: *Dialogue concerning the two chief world systems: Ptolemaic and Copernican*, trans. Stillman Drake (New York: The Modern Library, 2001), pp. 216–218.

[80]For making this point clear and enlightening the role played by the medioeval philosophy in the incubation of the scientific revolution of the XVII century, we refer to the work of Edward Grant: *The Foundation of Modern Science in the Middle Ages. Their Religious, Institutional and Intellectual Contexts* (Cambridge University Press, 1996).

experimental measurements on the Earth were carried out. A method implemented with success was that due to the Islamic mathematician and astronomer Abu Arrayhan ibn Ahmad al-Biruni (973–1048), who used trigonometric calculations for measuring the circumference of the Earth by sighting the horizon from the top of a mountain of known height.

Since this point is beyond our (Western!) interest we refer the reader elsewhere.[81]

Suggested Readings

Borst, A. (1993). *The ordering of time: From the acient computus to the modern computer* (1st German ed., 1990). University of Chicago Press.

Colish, M. L. (1997). *Medieval foundations of the western intellectual tradition*. Yale University Press.

Curtius, E. R. (2013). *Literature and the Latin middle ages* (1st German ed., 1948). Princeton University Press.

Freely, J. (2010). Aladdin's lamp: How greek science came to europe through the islamic world. Vintage Books Edition.

Haskins, C. H. (1927). *The renaissance of the twelfth century* (rpt. The World Publishing Company, 1957). Harvard University Press.

Kieckhefer, R. (2014). *Magic in the middle ages* (2nd ed.). Cambridge University Press.

Linberg, D. C. (Ed.). (1978). *Science in the middle ages*. University of Chicago Press.

Linberg, D. C. (1992). *The beginning of western science: The european scientific tradition in philosophical, religious, and institutional context, prehistory to A.D. 1450*. University of Chicago Press.

[81]Amelia C. Sparavigna: "The Science of al-Biruni", *International Journal of Sciences* 2(12) (2013): 52–60; Lawrence D'Antonio: "How to Measure the Earth", in D. Jardine & A. Shell-Gellasch (eds.), *Mathematical Time Capsules: Historical Modules for the Mathematics Classroom*, pp. 7–16. Mathematical Association of America.

Chapter 4
From the Age of the Great Transoceanic Discoveries to the New Measurements of the Earth

The last decades of the fifteenth century and a great part of the sixteenth represent the age of the great transoceanic navigations and of the consequent new geographic knowledge. From the discovery of America (Columbus) to the great circumnavigation of the world (Magellan), the great enterprises of the navigators confirmed the sphericity of the Earth, while the measurements of the meridian carried out on the ground updated the measure of its size. We shall deal with the obtained results by considering the works of the authors who both suggested and affected, and narrated the fundamental enterprises.

The fifteenth century saw the first great undertakings of the Portuguese navigators, followed by Columbus' discovery of America, but the middle of the century also witnessed the invention of the art of printing. As we shall see, the two things would end up intersecting. Some newly printed books were to have a great influence on the transoceanic discoveries. In any case, following a personal predilection, we first want to make an observation regarding the first books printed in Italy.

As is known, the two printers (and German monks) Konrad Schweinheim (or Sweynheym) and Arnold Pannartz went to Italy in 1464, and installed a printing house at Subiaco, in the Benedectine monastery of St. Scholastica, which was mostly occupied by German-speaking monks linked to the monastery of Aufsburg. From our point of view, it is interesting to see what first works printed were. In chronological order, they were:

1. *Donatus pro puerulis* (a Latin grammar for children; recall that the grammarian Aelius Donatus had been the master of S. Jerome);
2. Cicero's *De Oratore*;
3. Lactantius's *Divinae Institutiones*. This is the oldest book printed in Italy, bearing the date of its publication, 29 October 1465; it was printed in 275 copies.
4. Augustine's *De Civitate Dei*, also printed in 275 copies.

The fact that, at a distance of time of about ten centuries from its composition, the work of Lactantius was inserted among the first to be printed, together with the works of S. Augustine, leads us to think that Lactantius' ideas were still dominant

D. Boccaletti, *The Shape and Size of the Earth*,
https://doi.org/10.1007/978-3-319-90593-8_4

in the milieu of the monasteries, and consequently among the faithful. In any case, the fact that Voltaire, three centuries later, in his *Dictionnaire philosophique*, at the entry "*Le ciel des anciens*", quotes Lactantius together with Augustine, deriding them with regard to their denial of the sphericity of the Earth, is significant, because it shows precisely that those ideas, at least among the uneducated people, had survived a long time.

Let us report a brief excerpt of Voltaire's text:

> St. Austin calls the notion of antipodes an absurdity; and Lactantius flatly says, "Are there any so foolish as to believe there are men whose head is lower than their feet?" St. Chrysostom, in his fourteenth homily, calls out, "Where are they who say the heavens are moveable, and their form round? Lactantius again says, b. iii, of his Institutions, "I could prove to you by a multitude of arguments, that it is impossible the heavens should encompass the earth". The author of Spectacle de la Nature is welcome to tell the chevalier over and over, that Lactantius and Chrysostom were eminent philosophers; still it will be answered that they were great saints, which they may be without any acquaintance with astronomy. We believe them to be in heaven, but own that in what part of the heavens they are we know not.[1]

As we shall see, other important books were published in the last quarter of the fifteenth century.

4.1 The Translation of Ptolemy's *Geography* and the Books that Columbus Annotated

Perhaps our idea concerning the influence that the work of Lactantius might have had on popular belief is not shared by everybody, but the recognition of the influence that some books exerted on the designs of the navigators of the fifteenth century is generally agreed upon. Before quoting and commenting on the most important of them, let us recall which were fundamentally, at the beginning of the century, the ideas on the extent of the oecumene. As the outcome of the inquiries of the philosophers of the Middle Ages, substantially two theories concerning the Earth and its size were in the field (we are speaking only of theories, since no measurement had been repeated in the West in the Middle Ages). According to the

[1]Voltaire: *Dictionnaire Philosophique Portatif* (London, 1764), pp. 63–64: *Aussi St. Augustin traite l'idée des Antipodes, d'absurdité; & Lactance dit expressément: Y a-t'il des gens assez fous pour croire qu'il y ait des hommes dont la tête soit plus basse que les pieds? Saint Chrysostôme s'écrie dans sa quatorzieme homélie, Où sont ceux qui prétendent que les cieux sont mobiles, & que leur forme est circulaires? Lactance dit encore, au Liv. III. de ses institutions; Je pourrais vous prouver par beaucoup d'arguments, qu'il est impossible que le ciel entoure la terre. L'Auteur du Spectacle de la Nature pourra dire à Monsieur le Chevalier tant qu'il voudra, que Lactance & saint Chrysostôme étaient de grands philosophes, on lui répondra qu'ils étaient de grands Saints, & qu'il n'est point du tout nécessaire pour être un Saint, d'être un bon astronome. On croira qu'ils sont au Ciel, mais on avouera qu'on ne sait pas dans quelle partie du Ciel précisément* (Eng. trans. *The philosophical Dictionary* (London, 1802), pp. 190–191.

first one, the oecumene was of great extent and consequently one assumed a relatively small distance between Europe and Asia via the Atlantic Ocean. The second held exactly the contrary. Both theories can be traced back to Aristotle.

As to the first, we can look to Aristotle, who in *On the Heavens* says:

> Hence one should not be sure of the incredibility of the view of those who conceive that there is continuity between the parts about the pillars of Hercules and the parts about India, and that in this way the ocean is one.[2]

After Aristotle, we also find Seneca, who in his *Naturales Quaestiones* says:

> Then contempt for the narrow limits of its former dwelling succeeds. For what after all is the space that lies from India to the farthest shores of Spain? A few days journey if a prosperous wind waft the vessel.[3]

In addition, in his tragedy *Medea* Seneca predicted:

> There will come an age in the far-off years when Ocean shall unloose the bonds of things, when the whole broad earth shall be revealed, when Tethys shall disclose new worlds and Thule not be the limit of the lands.[4]

Cesare De Lollis, one of the most passionate biographers of Christopher Columbus, wrote:

> And these beautiful verses, where the prophecy is tuned to the unshakable certainty of a biblical vision, were held in high repute by all the cosmographers of the fifteenth century, and more than once occur under the pen of Columbus.[5]

In the Middle Ages, Aristotle's opinion was reprised by Roger Bacon, who in his *Opus Maius*, speaks about a "*parvum mare*", says:

> This is why he concludes that these places are not far apart, and there must be only a little sea between them. The sea, then, does not cover three quarters of the earth, as it has been guessed.[6]

The second theory is the model of the four elements arranged in concentric spheres (with the supplement of the medieval lucubrations on the inhabited lands) and provides for a small oecumene emerging from the predominant sphere of water.

---

[2]Aristotle: *On the Heavens*, op. cit., II, 14, III (Eng. trans. J. L. Stocks).

[3]Seneca: *Naturales Quaestiones*, I, pref. 13: *Quantum enim est, quod ab ultimis littoribus Hispaniae usque ad Indos iacet? Paucissimorum dierum spatium, si nauem suus uentus impleuit.* (Eng. trans. John Clarke).

[4]Seneca: *Medea*, 375–379: *Uenient annis saecula seris, /quibus Oceanus uincula rerum/laxet et ingens pateat tellus/Tethysque nouos detegat orbes/nec sit terris ultima Thule*" (Eng. trans. Frank Justus Miller).

[5]Cesare De Lollis: *Cristoforo Colombo nella leggenda e nella storia*, 3rd ed. (Treves 1923) (our Eng. trans.)

[6]Roger Bacon: *Opus Maius*: *Quapropter concludit haec loca esse propinquiora, et ideo oportet quod mare sit parvum inter ea. Non igitur mare cooperiet tres quartas terrae, ut aestimatur* (Eng. trans. H. M. Howe).

Therefore, the expanse of water between Europe and Asia, from orient to occident, could not have been covered by the ships in operation in the fifteenth century.

By the middle of the century Portuguese navigators had sailed along the western coast of Africa and had pushed farther and farther southward, proving that it was possible to pass the Equator without being incinerated. These voyages also drove them to the discovery of the islands facing the western coast of Africa. The Portuguese, who at that time were the most important experienced navigators, had as their goal to reach India by going around Africa and, as we know, Vasco da Gama achieved that goal in 1488, rounding the Cape of Good Hope. King Afonso V of Portugal was also interested in ascertaining if the Atlantic crossing had been possible (again, obviously, for reaching India), and for this reason he sought advice from the Italian scientist Paolo dal Pozzo Toscanelli (1397–1482). Toscanelli replied (on 25 June 1474) by sending to Fernão Martins, canon of Lisbon, a letter (which later on became famous) for the king, in which he maintained it was possible to reach the Indies by starting from the islands of the Atlantic Ocean already discovered by the Portuguese navigators (we will come back to this letter below).

It is not our intention to deal with the sea voyages undertaken in the fifteenth century by the navigators from various countries (Italy, Portugal, etc.) and the maps they used. That would indeed be a "*mare magnum*" in which we could not sail. Our only purpose is that of establishing how those navigators improved the knowledge of the Earth and its size and, as a consequence, knowledge of the extent of the inhabited lands. Thus, as anticipated in the title of this section, we shall deal with some books which also began to attract the interest of navigators.

As we have said, in the ancient world Pytheas of Massalia described his voyage through the Atlantic Ocean, sailing around Europe up to the extreme North (the mythic island of Thule), but nobody after him (at least as far as we know) had done the same.

In the Middle Ages, the travellers' tales concerned travels effected mostly overland, by merchants or by pilgrims heading for the Holy Land. One could say that in most, or, rather, in all cases, the tales were embellished with quite imaginary events or places. Therefore, to extract geographic knowledge from these tales often implied making mistakes. We can consider, as examples, two of these tales which rose to fame in Europe for several centuries, were translated into various languages, and were also appreciated for their literary quality.

Let us begin from the most ancient: *The Travels of Marco Polo* (also known as *The Million*).[7] Marco Polo was a merchant of a Venetian family who set off for the East together with his father and his uncle and stayed there for 25 years, most of which were spent in China. After his return, he took part in the war between the Republic of Venice and that of Genoa and was taken prisoner by the Genoese. In prison, in 1298, in Genoa, he shared a cell with Rustichello da Pisa, an author of romances taken prisoner during a battle of Pisa against Genoa, to whom he dicated

[7]See, for instance, Marco Polo: *The Travels*, trans. Ronald Latham (New York: Penguin, 1958).

his book. The book, conceived as a description of the world, that is, as a geographical treatise in which Marco only intervened to assert the veracity of the narration, was originally written in French and, almost certainly, its title in ancient French was *Le divisament dou monde*, i.e., the description of the world.

As we have said above, in works of this type, imaginary events and protagonists made their appearance. Even Marco indulged in this practice, though only to a certain degree. In his book, we find a figure who appears in all narrations of this type: Prester John. He was a legendary Christian ruler of the East, popularized in medieval chronicles since the period of the Crusades, to whom many military victories against the Muslims were ascribed. In addition, every teller talked about him as if he was living and protagonist of episodes in the age of the writer.

We meet Prester John again (more than sixty years after) in the second book we have chosen to take into consideration, *The Travels of Sir John Mandeville*,[8] written in 1363 in French by an Englishman. According to modern scholars, most of the book must be considered a work of compilation rather than a report of travels actually carried out, although attempts have been made to identify the sources. In any case, since it met a great success among the contemporaries, both navigators and cosmographers, let us consider some of its assertions.

In first place, it can be seen that Mandeville believes in the sphericity of the Earth. In fact, he says:

> In that land, ne in many other beyond that, no man may see the Star Transmontane, that is clept the Star of the Sea, that is unmovable and that is toward the north, that we clepe the Lode-star. But men see another star, the contrary to him, that is toward the south, that is clept Antartic. And right as the ship-men take their advice here and govern them by the Lode-star, right so do ship-men beyond those parts by the star of the south, the which star appeareth not to us. And this star that is toward the north, that we clepe the Lode-star, ne appeareth not to them. For which cause men may well perceive, that the land and the sea be of round shape and form; for the part of the firmament sheweth in one country that sheweth not in another country. And men may well prove by experience and subtle compassment of wit, that if a man found passages by ships that would go to search the world, men might go by ship all about the world and above and beneath.[9]

But, as Randles ironically comments,[10] the concept of Earth's sphericity alleged by Mandeville would have been more appreciated by Lactantius than by Aristotle:

> And wit well, that, after that that I may perceive and comprehend, the lands of Prester John, Emperor of Ind, be under us. For in going from Scotland or from England toward Jerusalem men go upward always. For our land is in the low part of the earth toward the west, and the land of Prester John is in the low part of the earth toward the east. And [they] have there the day when we have the night; and also, high to the contrary, they have the night when we have the day. For the earth and the sea be of round form and shape, as I have said before; and that that men go upward to one coast, men go downward to another coast. Also ye have heard me say that Jerusalem is in the midst of the world. And that may men prove, and

[8]See: *The Travels of Sir John Mandeville: The version of the Cotton Manuscript in modern spelling* (New York: Macmillan, 1900).

[9]*The Travels of Sir John Mandeville*, op. cit. p. 120.

[10]W. G. L. Randles: *De la terre plate*, op. cit., Chap. I, p. 6.

> shew there by a spear, that is pight into the earth, upon the hour of midday, when it is equinox, that sheweth no shadow on no side. And that it should be in the midst of the world, David witnesseth it in the Psalter, where he saith, Deus operatus est salutem in medio terrae. Then, they, that part from those parts of the west for to go toward Jerusalem, as many journeys as they go upward for to go thither, in as many journeys may they go from Jerusalem unto other confines of the superficiality of the earth beyond. And when men go beyond those journeys toward Ind and to the foreign isles, all is environing the roundness of the earth and of the sea under our countries on this half.[11]

Nevertheless (as Randles further comments), Mandeville does not follow Lactantius on the problem of Antipodes:

> But how it seemeth to simple men unlearned, that men ne may not go under the earth, and also that men should fall toward the heaven from under. But that may not be, upon less than we may fall toward heaven from the earth where we be. … For if a man might fall from the earth unto the firmament, by greater reason the earth and the sea that be so great and so heavy should fall to the firmament: but that may not be, and therefore saith our Lord God, Non timeas me, qui suspendi terram ex nihilo?[12]

Summarizing, Mandeville's compilation merges the ancient tradition and the narration of the Bible, by replacing the ancient tradition of the well in Syene with a spear fixed in the ground in Jerusalem, which is considered the centre of the world, but mistaking the equinox for the solstice.

In spite of all, both works had numerous printed editions in the second half of the fifteenth century and were known and appreciated by the most famous navigators as well as cosmographers: suffice it to mention Henry the Navigator, Christopher Columbus and Paolo dal Pozzo Toscanelli.

As we have already said in Sect. 1.8.2, Ptolemy's *Geography* (in Latin translation) was also printed in the last decades of the fifteenth century. To be precise, the first edition (without maps) appeared in Vicenza in 1475, followed by a second one in Bologna, with maps, in 1477. All told there were six editions before 1500. In the same years there were numerous printed editions of both *The Million* and *The Travels of Sir John Mandeville*.

Unlike these two "descriptive" works, the appearance of Ptolemy's work put mathematics once again into the field as a tool for representing the oecumene on a map. The projections of Ptolemy, with their representation of the meridians and the parallels, became necessary when the oceans were navigated and mariners depended on the aid of the stars to orient themselves. The Portuguese were the first to use these charts,[13] when they had to navigate far from the coasts which were taken as landmarks.

Obviously, the use of Ptolemy's *Geography* had as a consequence that the Earth was considered as having a continuous spherical surface where earth alternated with water, in contrast to the Aristotelian conception. As we know, the oecumene

---

[11]*The Travels of Sir John Mandeville*, op. cit. p. 122.

[12]Ibidem, p. 123.

[13]See Johannes Keuning: "The History of Geographical Map Projections until 1600", *Imago Mundi*, vol. XII (1955), p. 15.

represented by Ptolemy had an extension of only 180° in longitude starting from the Canaries. Therefore, his map could not be used for planning travels westward in search for the coasts of the Indies.

It seems instead (but on this subject many are the legends) that another map affected the planning of the sailing of the Atlantic Ocean: the *one* that Paolo dal Pozzo Toscanelli would have sent to King Afonso V together with the famous letter we have mentioned above. We recall that Toscanelli, who studied mathematics and medicine in Padua and then did his intellectual work in Florence, was very renowned and esteemed in the intellectual milieu of Florence and throughout Europe. He was a friend of Filippo Brunelleschi, Leon Battista Alberti, Nicholas of Cusa, and Regiomontanus (Johannes Müller of Königsberg) and enjoyed a general reputation as a mathematician and cosmographer. Thus, it was quite natural that the king of Portugal addressed him for shedding some light on the possibility of reaching the Indies by crossing the Atlantic Ocean. Toscanelli, in his letter, speaks of a map he himself had drawn which showed the islands from which one must start and the places (today's Japan?) where one would arrive by sailing continuously westward. This letter, evidently, did not convince King Afonso to finance a fleet to that end (nor his successor João II, who upheld the same line[14] even when in 1483–84 Columbus himself proposed the undertaking), but enthralled Columbus, who had received a copy of it from Toscanelli himself.

What we are trying to do is to show to what extent the interaction between the cultural world and the world of the navigators, and of the courts which sponsored their enterprises, has caused new knowledge about the inhabited world and the Earth in general. Until now, we have seen what works of the past concerning the travels or the description of the world were made available by the nascent publishing industry. Now let us take into consideration two works published in the last quarter of the fifteenth century which seem to have been much consulted by Christopher Columbus. Since the historians who have dealt with Columbus had the opportunity of carrying out a survey of his personal library and even of transcribing the margin notes personally made by him on those texts, we are able to know which works he held in high regard. This is a possible criterion for appreciating which works of that period were considered authoritative by the navigators who had to undertake oceanic voyages. That is, we consider Columbus to be an indicator of the interests of the navigators of that time.

The works most annotated by Columbus are two: *Imago Mundi*[15] by Pierre D'Ailly and *Historia rerum ubique gestarum*[16] by Aeneas Sylvius Piccolomini (later Pope Pius II). The first modern scholars who have dealt with the work of

[14]On this, see W. G. L. Randles: The Evaluation of Columbus' 'India' Project by Portuguese and Spanish Cosmographers in the Light of the Geographical Science of the Period", *Imago Mundi*, vol. 42 (1990) pp. 50–64.

[15]See: Edmond Buron, ed.: *Ymago Mundi de Pierre D'Ailly cardinal de Cambrai et chancelier de l'Université de Paris* (Paris, Maisonneuve Freres, 1930). There is also a partial English translation by Edward Grant in *A Source Book in Medieval Science*, op. cit., pp. 630–639.

[16]Aeneas Sylvius Piccolomini: *Historia rerum ubique gestarum* (Venice 1477).

Pierre D'Ailly also have drawn up a sort of table of the margin notes written by Columbus on the texts he had consulted[17]: there are 898 notes on D'Ailly's work, and 861 on *Historia rerum ubique gestarum*. After these, the most annotated works are those of Marco Polo with 366, and Pliny with 24. The heavily annotated *Imago Mundi* has even been called Columbus' "bedside companion".

At this point, one would expect the *Imago Mundi* to be a new and original work (obviously, for that time), instead we have to do with a compilation, a sort of updated Isidore, though he only quotes ancient books. In fact Louis Salembier says:

> *Le cosmographe cambrésien a le tort de ne point citer ses contemporains, qu'il a l'air de vouloir ignorer de parti pris. Toute sa science est "livresque", comme dira plus tard Montaigne, et tout ce qui n'a pas été écrit a plusieurs siècles de distance semble ne point exister pour lui. Les traditions historiques écrites ou orales, que Bacon avait reproduites, paraissent lui avoir été complètement inconnues.*[18]

Columbus' margin notes are often limited to repeat some words of the text by pointing out them, but in some cases even update the knowledge supplied by the author. For instance, margin note 16 in Chap. 6 corrects the text where the author maintains that the so-called torrid zone is uninhabitable (*Ideo vocatur zona torrida et dicitur inhabitabilis propter nimium calorem*). Columbus writes: "The Torrid zone is not uninhabitable, because today the Portuguese navigate through it; indeed it is very populated".[19]

Another example of this type is found at the beginning of chap. VIII, where the author quotes the opinions of Ptolemy and Aristotle on the inhabitability of the lands straddling the Equator. Columbus, in his margin note, reports the enterprise of Bartolomeu Dias who, with three caravels sent to Guinea by King João II, arrived in 1488 at the Cape of Good Hope. What one can conclude is that Columbus, while greatly interested in the arguments (all based on the authority of the classics) furnished by D'Ailly, which he would use for convincing others about the feasibility of his plan to sail westward and reach the Orient, was also (as a navigator) careful to take note of the news from the sea-going enterprises of his contemporaries.

As has been remarked by scholars, D'Ailly, besides drawing from the classics, also and abundantly drew from a work in French (*Sphere*) by Nicole Oresme without naming its author, and, of course, not realizing the contradictions among the different authors he copied from.

---

[17]See: Lous Salembier: "Pierre D'Ailly et la découverte de l'Amerique", *Revue d'histoire de l'Église de France*, tome 3, n° 16 (1912?, pp. 377–396; see also E. Buron, *Ymago Mundi de Pierre D'Ailly*, op. cit., p. 27.

[18]L. Salembier, loc. cit. p. 386. Buron's comment is even more caustic; see E. Buron, ed.: *Ymago Mundi*, op. cit., p. 152, footnote 3; a English translation of this is found in E. Grant, *Source Book in Medieval Science*, op. cit. p. 630.

[19]Edward Grant, *A Source Book in Medieval Science*, op. cit. p. 635.

More or less the same opinion can be expressed about the work of Aeneas Silvius Piccolomini (1405–1464). As is known, Piccolomini, besides having had a career in the Church, before being elected Pope had written several literary works and had planned to narrate the histories of all peoples of the Earth through the description of their places in the three parts of the world known at that time, Asia, Europe, and Africa. The work was supposed to have been entitled *Historia rerum ubique gestarum locorumque descriptio* and was intended to be a history of human existences in all places, but intended to be at the same time a cosmography including a historical-geographical encyclopaedia. The work remained unfinished (the part concerning Africa was not written) and the parts regarding Asia and Europe were published after his death in 1477 with the title *Historia rerum ubique gestarum*[20]; this is the edition annotated by Columbus. As we have said above, this work has exactly the same defects of *Imago Mundi*, which W. G. L. Randles summarises like this: "At no point in his generous eclecticism, does Piccolomini make a critical choice between the conflicting theories he reviews, not even to choosing between Homer and Ptolemy".[21]

Thus far we have spoken of these works (of D'Ailly and Piccolomini) as actually annotated by Columbus and as if the margin notes were made prior to his enterprises and then used as elements in his project. But many scholars have expressed doubts about this, some maintaining that the notes were appended after the travels, others that they were even written by his brother Bartolomew. In any case, these discussions do not alter the conclusion that navigators at that time, in addition to the available maps, also consulted the works of a cosmographic nature, both the recent encyclopedias and the works of the ancient philosophers. Although the increasing number of ocean voyages favoured the writing of new works, some of them were still tied to the old conceptions[22] (in the sense that, while the fact of a spherical Earth was accepted, the question of the proportion between the waters and the dry land remained open).

The work that (at least from a certain point of view) we could define as conclusive is credited to Amerigo Vespucci: *Mundus novus*. This work has also been much discussed. Firstly, at least two versions of it exist. The first, which appeared perhaps in Paris in about 1503, was a description in Latin of the 1501 voyage, allegedly written by Vespucci. Translations of it into various languages became best-sellers all over Europe. The other text is known as the *Lettera di Amerigo Vespucci a Lorenzo di Pierfrancesco dei Medici*, published in Florence about 1505–1506 and describing all real and alleged voyages of Vespucci. The conclusions drawn by Vespucci were two: (1) that it was possible to pass the Equator going southward and that there the land was inhabited and abounding in animals; (2) that

[20]There is a recent critical edition of the part regarding Asia; see Enea Silvio Piccolomini, Papa Pio II, *Asia*, Nicola Casella, ed. (Bellinzona: Edizioni Casagrande, 2004).

[21]See W. G. L. Randles, *De la terre plate*, op. cit., p. 38.

[22]See W. G. L. Randles, *De la terre plate*, op. cit., pp. 43–53.

that land was not India but a new continent (*mundus novus*). In the end, what we could call an "experimental test" had prevailed over the endless theoretical discussions.

4.2 Jean Fernel and the New Measure of the Degree of the Terrestrial Meridian

While on the one hand the diffusion of Vespucci's booklet put an end to the idea that the lands discovered by Columbus were the Indies, it did not on the other hand stop the speculations about the proportion between the water and the dry land, even if nobody any longer believed the Earth to be a ball immersed in the water. Among the works published in the first decades of the sixteenth century, one assumes a particular importance: the *Cosmotheoria*[23] of Jean Fernel.

Fernel was a French physician, mathematician and astronomer (Montdidier, 1497–Fontainebleau, 1558). After studies of physics and mathematics, he devoted himself to the study and practice of medicine, and in 1556 became the archiater (chief physician) of King Henry II of France. In the field of medicine he obtained important results and wrote famous works, but we are interested in the *Cosmotheoria* mentioned above.

In it, Fernel, after having rejected the theory of Albert of Saxony and Buridan of the two distinct centres (one of size and one of gravity), explains his own theory (which he traces back to Aristotle) according to which the surface of the dry land and that of the seas constitutes a unique spherical surface:

> In the first place all philosophers, particularly Aristotle with regard to heavens (and this for maximally natural reasons) maintain that the surface of the inhabited land constitutes, with the water washing it, a unique and convex surface, whose centre is the same of the centre of the universe. In fact it appears evident from both the writings of the scholars and the faithful testimony of the navigators that the sea itself is here and there covered by innumerable islands and several shoals, which retain the same convex shape of the sea, and all agree that these places are no less distant from the centre of the universe than the surface of the inhabited land.[24]

As one can see, Fernel, to corroborate his conclusions, did not limit himself to relying on Aristotle (the customary resort to the classics) but also quoted "the

[23]Joannis Fernelii Ambianatis (Jean Fernel): *Cosmotheoria, libros duos complexa-Parisiis* (Paris: in aedibus Simonis Colinaei, 1528).

[24]Joannis Fernelii Ambianatis (Jean Fernel): *Cosmotheoria*, I, cap. I, 6: *Probant in primis philosophi omnes, Aristoteles praesertim secundo coeli idque rationibus quae maxime naturales sunt, terrae faciem habitatam, una cum aquae connexo, superficiem unicam, eamque convexam efficiere, cuius et universi idem sit centrum. Quum igitur perspicuum sit tum eruditorum virorum monimentis, tum fideli navigantium testimonio, mare ipsum innumeris insulis, plurimisque syrtibus passim cospersum esse, quaedam fere cum mari convexum retinent, consentaneum est et ea loca non minus ab universi centro removeri, quam hanc habitatam terrae faciem* (our Eng. trans.).

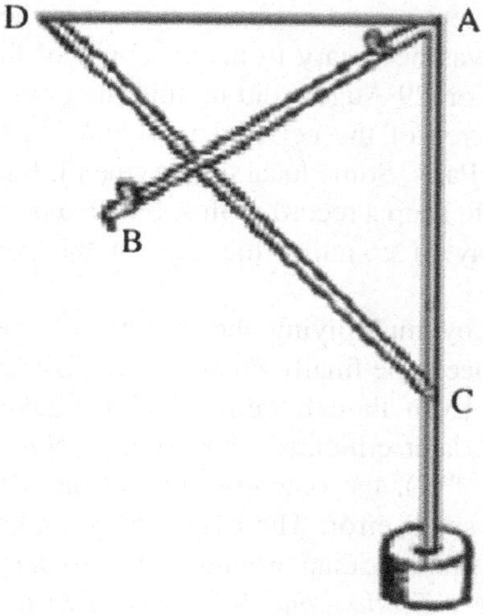

Fig. 4.1 Fernel's triquetrum for measuring the altitude of the noonday sun

faithful testimony of the navigators". Finally, we see a resort to "experimental proofs"!

From the point of view of the shape, he writes:

> Therefore, one must recognize that the Earth has the appearance of a wooden globe in which there are really a lot of concavities, where the water can be received.[25]

Thus, in speaking of a globe with cavities which contains the seas, he reached a correct representation of the Earth.

But this is not all: Fernel also supplied the measure of its size. At a distance of eighteen centuries from Eratosthenes, Fernel realized the first measure of the degree of the terrestrial meridian in the Western world in modern times. As we shall see, his method was not extremely rigorous, even if more precise (at least for the land distances) than the measures carried out in Egypt at Eratosthenes' times by the bematists. In any case, the opinion of Pierre Duhem was a pithy judgement: "… *il eut le bonheur d'obtenir un résultat exact par un procédé qui l'état fort peu*".[26]

The measurement began on 25 August 1527, in Paris. Fernel, with the use of the triquetrum (see Fig. 4.1),[27] measured the altitude of the noonday Sun. Starting from Paris and going north on foot (on the road for Amiens), what he had to do was to find the point where the altitude of the noonday Sun was such to confirm that one had covered one degree of the celestial meridian.

---

[25]Joannis Fernelii Ambianatis (Jean Fernel): *Cosmotheoria* I, cap. I, 7: *Proinde existimandum est, terram globi cuiusquam lignei speciem habere, in quo concavitates plurimae sunt, quibus aqua recepi possit* (our Eng. trans.).

[26]P. Duhem: *Les origines de la Statique*, tome II, 1906, p. 348.

[27]The triquetrum was the medieval name of an astronomical instrument used for determining the altitude of heavenly bodies (already known to Ptolemy).

Since these measurements could not be performed at the same time, but instead at intervals of days, it was necessary to take account of the daily decrease of the Sun's declination. Only on 29 August did he find the point which corresponded to the decrease of one degree of the celestial meridian. At that point he needed to know the distance from Paris. Some local countrymen informed him that they were 25 leagues from Paris. He keep a record of this, but also boarded the stage coach for Paris and devised a way of counting the laps of the wheels, of which he had measured the diameter.

On arrival at Paris, by multiplying the number of the laps (17,024) for the circumference of the wheels, he finally obtained the distance covered (a little more than 110 current km). Even though Fernel had not taken account of spherical trigonometry (Delambre later criticized him, saying "*N'avait-il aucune idée de la trigonométrie sphérique?*"[28]), the outcome for the measure of the terrestrial circumference had a very small error. The comment of Delambre, who, as we shall see, measured the terrestrial meridian around 270 years later, was: "*C'est un grand bonheur; c'est la seule reflection que je me permettrai sur la mesure de mon compatriote*".[29]

After Fernel, in the sixteenth century there were several authors of works regarding the shape and the structure of the Earth. Just as Aristotle's cosmology had taken time in the Middle Ages for gaining ground, now in the Renaissance it took time to be abandoned. Therefore in the same period works suggesting different models for the structure of the Earth coexisted. At this point the spherical shape was generally agreed upon, but the debate on the placement of the water persisted. One work which met a considerable success and enjoined many reprints is due to an author with a very distinguished family name at that time: Alessandro Piccolomini (1508–1578).[30]

Before talking about the work, let us spend a few words about the author. After an initial literary activity in Siena, Piccolomini moved to Padua where he studied mathematics and astronomy at the University. There he wrote his first "scientific" works: *De la sfera del mondo* (Venice, 1540) and *De le stelle fisse* (Venice, 1540). Afterwards, he moved to Bologna, Siena and Rome (where he took holy orders) and finally back to Siena, where he wrote the work we alluded to above: *Della grandezza della terra et dell'acqua* (Venice, 1558). As one can see from the titles, the three works we have quoted are not written in Latin, but in Italian. Probably due to his friendship, in Padua, with the man of letters Sperone Speroni, who promoted the use of the Italian, or perhaps to allow them to be read by a gentlewoman friend of his who did not know Latin, the fact is that Piccolomini preceded Galileo in publishing scientific works in Italian. The value of these works lies not so much in

[28]Jean-Baptiste Delambre: *Historie de l'Astronomie du Moyen Âge* (Paris, 1819), p. 383.

[29]Ibidem, p. 385.

[30]See Simone Mammola: *Il problema della grandezza della terra e dell'acqua, negli scritti di Alessandro Piccolomini, Antonio Braga e G. B. Benedetti e la progressiva dissoluzione della cosmologia delle sfere elementari nel secondo '500*, Preprint 459, Max Planck Institute for the History of Science, 2014.

the scientific novelties as in the author's ability to make scientific matters accessible for a larger readership.

In *Della grandezza della terra et dell'acqua*, Piccolomini follows the traditional procedure of the medieval works by first exposing the arguments supporting the thesis of the superiority of the water over the land (ten arguments, five of which taken from the experience and five from the authorities). The authorities are the usual: Aristotle, Strabo, Pomponius Mela, Pliny. He then passes on to elaborate his reply to the explained arguments. First of all, he remarks that the question is twofold: the relative superiority or inferiority of the two elements can concern both the surface and the volume.

First, Piccolomini demonstrates that the land exceeds the water in surface. He does this through an "extremely original mixed mathematical-empirical proof".[31] After having obtained a globe (considered trustworthy), he drew on it a network of parallels and meridians duly interspersed in order to calculate their areas, proving in this way that the area of the land was much greater than that of the water. But it must be said that in the globes of that time the extension of the land was overestimated.

Piccolomini also succeeded in demonstrating the superiority of the volume and so demolishing the theory of the proportion 10:1 held by the Aristotelians. In this use of the network of parallels and meridians one can easily recognize the lesson of Ptolemy's *Geography*.

4.3 Nicolaus Copernicus and Christopher Clavius

The concept of terraqueous globe, which already emerges from the way in which Piccolomini formulates the problem, had already been definitely shared years before by Copernicus (1473–1543), who in the beginning of his work *De Revolutionibus orbium coelestium*[32] devotes the second and the third chapter of the first book to this subject. We report two excerpts (the short second chapter and the beginning of the third) to point out the argumentations on which Copernicus founded his conclusions:

> Chapter 2. The Earth is spherical
>
> The earth also is spherical, since it presses upon its center from every direction. Yet it is not immediately recognized as a perfect sphere on account of the great height of the mountains and depth of the valleys. They scarcely alter the general sphericity of the earth, however, as is clear from the following considerations. For a traveler going from any place toward the north, that pole of the daily rotation gradually climbs higher, while the opposite pole drops down an equal amount. More stars in the north are seen not to set, while in the south certain

[31]S. Mammola: *Il problema della grandezza*, op. cit., p. 29 (our trans.).

[32]Nicolai Copernici Torinensis (Nicolaus Copernicus): *De Revolutionibus orbium coelestium, Libri VI* (Nurimberg: apud Joh. Petreius, 1543).

stars are no longer seen to rise. Thus Italy does not see Canopus, which is visible in Egypt; and Italy does see the River's last star, which is unfamiliar to our area in the colder region. Such stars, conversely, move higher in the heavens for a traveler heading southward, while those which are high in our sky sink down. Meanwhile, moreover, the elevations of the poles have the same ratio everywhere to the portions of the earth that have been traversed. This happens on no other figure than the sphere. Hence the earth too is evidently enclosed between poles and is therefore spherical. Furthermore, evening eclipses of the sun and moon are not seen by easterners, nor morning eclipses by westerners, while those occurring in between are seen later by easterners but earlier by westerners. The waters press down into the same figure also, as sailors are aware, since land which is not seen from a ship is visible from the top of its mast. On the other hand, if a light is attached to the top of the mast, as the ship draws away from land, those who remain ashore see the light drop down gradually until it finally disappears, as though setting. Water, furthermore, being fluid by nature, manifestly always seeks the same lower levels as earth and pushes up from the shore no higher than its rise permits. Hence whatever land emerges out of the ocean is admittedly that much higher.[33]

This short chapter is substantially a summary of what was already known since antiquity and exposed in the first book of Ptolemy's *Almagest*. The third chapter, instead, even in the title, represents a stance in support of a "terraqueous globe":

Chapter 3. How earth forms a single sphere with water

Pouring forth its seas everywhere, then, the ocean envelops the earth and fills its deeper chasms. Both tend toward the same center because of their heaviness. Accordingly there had to be less water than land, to avoid having the water engulf the entire earth and to have the water recede from some portions of the land and from the many islands lying here and there, for the preservation of living creatures For what are the inhabited countries and the mainland itself but an island larger than the others? We should not heed certain peripatetics who declared that the entire body of water is ten times greater than all the land. For, according to the conjecture which they accepted, in the transmutation of the elements as one unit of earth dissolves, it becomes ten units of water. They also assert that the earth bulges

[33]Nicolai Copernici Torinensis (Nicolaus Copernicus): *De Revolutionibus: Quòd terra quoque sphaerica sit. Cap. II. Terram quoque globosam esse, quoniam ab omni parte centro suo innititur. Tametsi absolutus orbis non statim videatur, in tanta montium excelsitate, descensuque vallium, quae tamen universam terrae rotunditatem minime variant. Quod ita manifestum est. Nam ad Septentrionem undequaque commeantibus, vertex ille diurnae revolutionis paulatim attollitur, altero tantundem ex adverso subeunte, pluresque stellae circum Septentriones videntur non occidere, & in Austro quaedam amplius non oriri. Ita Canopum non cernit Italia, Ægypto patentem. Et Italia postremam fluvii stellam videt, quam regio nostra plagæ rigentioris ignorat. E contrario in Austrum transeuntibus attolluntur illa, residentibus iis, quæ nobis excelsa sunt. Interea & ipsæ polorum inclinationes ad emensa terrarum spacia eandem ubique rationem habent, quod in nulla alia quàm sphærica figura contingit. Unde manifestum est, terram quoque verticibus includi, et propter hoc globosam esse. Adde etiam, quòd defectus Solis et Lunæ vespertinos Orientis incolae non sentiunt: neque matutinos ad occasum habitantes: Medios autem, illi quidem tardius, hi vero citius vident. Eidem quoque formæ aquas inniti à navigantibus depræhenditur: quoniam quæ è navi terra non cernitur, ex summitate mali plerumque spectatur. At vicissim si quid in summitate mali fulgens adhibeatur, a terra promoto navigio, paulatim descendere videtur in littore manentibus, donec postremo quasi occiduum occultetur. Constat etiam aquas sua natura fluentes, inferiora semper petere, eadem quæ terra, nec à littore ad ulteriore niti, quàm convexitas ipsius patiatur. Quamobrem tanto excelsiorem terram esse convenit, quæcunque ex Oceano assurgit* (Eng. trans. Edward Rosen).

> out to some extent as it does because it is not of equal weight everywhere on account of its cavities, its center of gravity being different from its center of magnitude. But they err through ignorance of the art of geometry. For they do not realize that the water cannot be even seven times greater and still leave any part of the land dry, unless earth as a whole vacated the center of gravity and yielded that position to water, as if the latter were heavier than itself. For, spheres are to each other as the cubes of their diameters. Therefore, if earth were the eighth part to seven parts of water, earth's diameter could not be greater than the distance from [their joint] center to the circumference of the waters. So far are they from being as much as ten times greater [than the land]. Moreover, there is no difference between the earth's centers of gravity and magnitude. This can be established by the fact that from the ocean inward the curvature of the land does not mount steadily in a continuous rise. If it did, it would keep the sea water out completely and in no way permit the inland seas and such vast gulfs to intrude. Furthermore, the depth of the abyss would never stop increasing from the shore of the ocean outward, so that no island or reef or any form of land would be encountered by sailors on the longer voyages.[34]

We point out the peculiarity of the proof that the water:earth proportion cannot be 10:1, that is, since it cannot even be 7:1, with yet stronger reason cannot be 10:1, therefore the assertion of the Peripatetics does not have a leg to stand on.

About thirty years after Copernicus, the subject of the terraqueous globe was resumed by Christopher Clavius in his work (already talked about in Sect. 3.5.1) *In Sphaeram Ioannis De Sacro Bosco Commentarius*.[35] As we have already said, this work had several editions. From the very start the commentary to Sacrobosco was essentially an opportunity for Clavius to express his own ideas on cosmography while at the same time fulfil his academic responsibilities. The various editions, up to the last of 1611 (in the *Collected works* Mainz 1611–12, published one year

---

[34]Nicolai Copernici Torinensis (Nicolaus Copernicus): *De Revolutionibus: Quomodo terra cum aqua unum globum perficiat. Cap. III. Huic ergo circumfusus Oceanus maria passim profundens, decliviores eius descensus implet. Itaque minus esse aquarum quàm terrae oportebat, ne totam absorberet aqua tellurem, ambabus in idem centrum contendentibus gravitate sua, sed ut aliquas terrae partes animantium saluti relinqueret, atque tot hincinde patentes insulas. Nam et ipsa continens, terrarumque orbis, quid aliud est quàm insula maior cæteris? Nec audiendi sunt Peripateticorum quidam, qui universam aquam decies tota terra maiorem prodiderunt. Quod scilicet in transmutatione elementorum ex aliqua parte terrae, decem aquarum in resolutione flant, coniecturam accipientes, aiuntque terram quadantenus sic prominere, quòd non undequaque secundum gravitatem æquilibret cavernosa existens, atque aliud esse centrum gravitatis, aliud magnitudinis. Sed falluntur Geometrices artis ignorantia, nescientes quòd neque septies aqua potest esse maior, ut aliqua pars terræ siccaretur, nisi tota centrum gravitatis evacuaret, daretque locum aquis, tanquam se gravioribus. Quoniam sphaeræ ad se invicem in tripla ratione sunt suorum dimetientium. Si igitur septem partibus aquarum terra esset octava, diameter eius non posset esse maior, quàm quæ ex centro ad circumferentiam aquarum: tantum abest, ut etiam decies maior sit aqua. Quòd etiam nihil intersit inter centrum gravitatis terræ, et centrum magnitudinis eius: hinc accipi potest, quòd convexitas terræ ab oceano expaciata, non continuo semper intumescit abscessu, alioqui arceret quàm maxime aquas marinas, nec aliquo modo sineret interna maria, tamque vastos sinus irrumpere. Rursum à littore oceani non cessaret aucta semper profunditas abyssi, qua propter nec insula, nec scopulus, nec terrenum quidpiam occurreret navigantibus longius progressis* (Eng. trans. Edward Rosen).

[35]Christopheri Clavii Bambergensis ex Societate Iesu (Christopher Clavius): *In Sphaeram Ioannis De Sacro Bosco Commentarius* (Rome: Apud Victorium Helianum, 1570).

before his death), contained several additions and revisions, but the long section devoted to the expositions of the reasons for holding the idea of the terraqueeous globe, entitled *An ex terra et aqua unus fiat globus, hoc est, an horum elementarum connexa seperficies idem habeant centrum*, was remained practically unchanged from the edition of 1581[36] up to the one included in the *Collected works* of 1611–12.

Before going into the writing of Clavius, we feel it is our duty to devote a few words to Clavius himself who, as he is not so well known outside of the community of experts. The biographical data on the first part of his life are very scanty. It seems that his first biographer, Bernardino Baldi (1553–1617), who wrote about him while he was still alive, even asked him directly for information. However, the outcome has been rather limited.[37] Undoubtedly he was born in Bamberg, a German town of Franconia, on 25 March 1538, but his original German name is not assuredly established (perhaps Klaue). He went to Rome when he was young and first appears in the records of the Society of Jesus in 1555. Afterwards, he studied in Coimbra and finally in Rome starting in 1561, where he first studied both physics and metaphysics at the Collegio Romano and successively theology, since a Jesuit, whatever science practised, also had to know theology. He was ordained in 1564 but is only recorded as having professed the solemn vows in September 1576, when he was thirty-seven years old. His career as a teacher began as early as 1563, and he always taught mathematics and astronomy. According to Baldi, he was an autodidact in these scientific disciplines, and although it has been suggested that he was influenced by Pedro Nuñes (d. 1578, professor in Coimbra), this has never been proved.

Clavius was the author of several works of mathematics, among which an edition of Euclid's *Elements*, which was not so much a translation of Euclid's geometrical texts as a paraphrase and commentary, as was his treatise on Sacrobosco's *Sphaera*. He also wrote a treatise on gnomonics, performed various astronomical observations and was one of the main promoters of the Gregorian reform of the calendar. In his last years he also met Galileo and appreciated his telescope.[38]

Clavius was not only a distinguished professor of the Collegio Romano. Besides having been the organizer of the teaching of the scientific disciplines in all Jesuit schools, he also was a deeply respected and esteemed scientist in the last decades of the sixteenth century. Therefore the excerpts we shall report are beyond question indicative of the official science at the end of the century of the discoveries of the European navigators.

[36]Christopheri Clavii Bambergensis ex Societate Iesu (Christopher Clavius): *In Sphaeram Ioannis De Sacro Bosco Commentarius Nunc iterum ab ipso Auctore recognitus & multis ac variis locis loclupetatus* (Rome: Ex Officina Dominici Basae, 1581), pp. 117–134.

[37]See: Bernardino Baldi, *Le vite de' matematici*, Elio Nenci, ed. (Milan: Franco Angeli, 1998), pp. 558–577.

[38]On this and on his position about Copernicus' theory we refer the reader to the fundamental book by James M. Lattis: *Between Copernicus and Galileo*, op. cit.

In what follows we report two excerpts in which he denies the possibility of the existence of two different centres for the Earth, one of the size and the other of the gravity. Clavius bases his assertion on what was related by the navigators, and concludes that those who maintain the existence of two different centres are proposing a theory which is "against experience". Now assertions are no longer based on the Bible, but on the navigators:

> Then, since the authors of both opinions admit that the water is much larger than the earth, they must necessarily also admit that to every degree of sea surface corresponds a number of stadia or of miles greater than that corresponding to a degree of earth surface. In fact, the earth's orb is divided up into the same number of degrees of the water orb and, as usual, of the whole celestial circle. On this account, if the water is larger than earth, then the degrees of the water must be larger than those of the earth and consequently must contain more stadia or miles than them. But all navigators state the direct opposite, precisely those who many times could realize from experience that there are so many stadia or miles in one degree on the earth surface as there are in one degree on the sea surface.[39]
>
> …
>
> Besides, if according to them [the surface of] the water was not at the same distance from the centre of the universe as [the surface of] the earth, but was much higher, consequently a ship leaving a port would ascend and arriving at the same port would descend; and thus, under the action of an equal wind, it should more rapidly descend than ascend, but this is against experience.[40]

He then continues the disputation, holding that also at the Antipodes the distribution of the dry land and the sea is analogous. Figure 4.2, which illustrates the theory of the two centres together with that of Antipodes, and will appear in all editions, is curious. Clavius gives the following explanation[41]:

> But perhaps they would say (as some people with whom I have discussed replied to me) that our Antipodes and the islands are contained in the same circumference with all the earth, and the sea is lifted between any two islands forming a bulge. From where, if it flowed out, it would re-cover the whole earth, including that part of it which is at the

---

[39]Christopher Clavius: *In sphaeram…*, pp. 117–118: *Deinde, quia cum auctores utriufque fententiæ admittant, aqua multo effe maiorem ipfa terra, concedere etiam neceffario cogentur, plura ftadia, milliariaue cuilibet gradui fuperficiei maris, feu aquæ correfpondere, quàm cuilibet gradui terræ. Nam in tot gradus diuiditur orbis terrenus, in quot globus aqueus diftribuitur, quemadmodum fcilicet quilibet circulus cœleftis diudi folet. Quare fi aqua maior eft, quàm terra, oportet gradus aquas effe maiores gradibus terræ, ac proinde quiuis illorum plura ftadia, milliariaue continebit, quàm quilibet horum, Cuius oppofltum omnes Nautae afferunt, qui fe expertos fuiffe fepenumero reftantur, tot ftadia, uel milliaria comprehendere unum quemque gradum in fuperficie. terræ, quot in fuperficie maris* (our Eng. trans.).

[40]Christopher Clavius: *In sphaeram…*, p. 118: *Praeterea, cum aqua fecundum illos non equaliter diftet a cen-tro Vniuerfi, fed eleuetur mirum in modum, fequeretur, quod nauis exiens è portu quocunque afcenderet, & accedens ad eundem portum defcenderet, & fic, equali exiftente uento, uelocius ad portum defcenderet, quàm a portu afcederet, quod eft contra experientiam: immo nullo parto confiftere poffet nauis extra portum conftituta, quin fua fponte ad portum decurreret, cum omne graue deorfum tendat; quod tamen uerum non eft* (our Eng. trans.).

[41]In this excerpt, the few phrases that are underlined indicate those that did not appear in the 1581 edition but were in the 1611 edition.

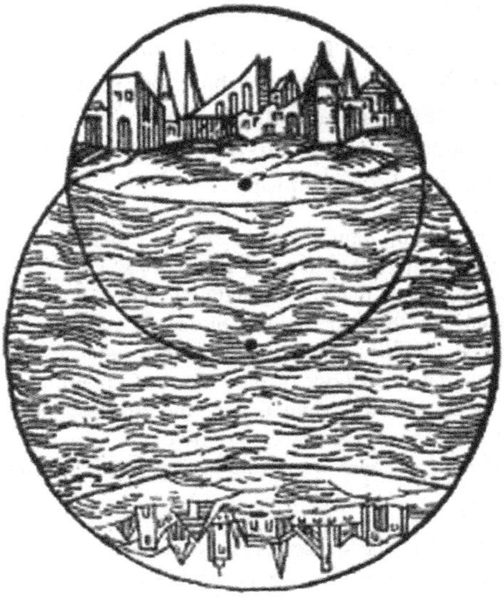

Fig. 4.2 The two centers of the world and the antipodes as depicted in Clavius' *In sphaeram*

> Antipodes together with all the islands. In truth, this reply is absurd. First, if it were that way, all the water would not have an only centre, but every bulge of water between two islands would have an its own centre, what is against the opinion of everyone and appears inconsiderately asserted.[42]

Finally, we report two more excerpts in which Clavius also bases himself on the authority of Fernel and Piccolomini:

> To this objection one must report that it proceeds from an untrue hypothesis; in fact it considers the earth to be only on one side and on the opposite side there is entirely sea, which is false. In fact, thanks to the navigations of our present time, one has discovered continents, islands or peninsulas under both the poles and the equinoctial circle, in both the East and the West, and finally in all the world, so that earth and water appear almost mixed in all the globe. In fact, the sea is dotted by an almost incalculable number of islands, so that it appears more covered by emersed lands than by waters, as Alexander Piccolomini perfectly proves in his booklet about the size of the earth and the water. Therefore, we say that this globe, which we assert to be made up of earth and water, is organized in such a way that the earth is superelevated all over the world and the water arranges itself in the

[42]Christopher Clavius: *In sphaeram…*, pp. 118–119: *Sed dicent fortaſſe, (ut aliqui mihi cum il-lis diſputanti reſponderunt) antipodes noſtros, & inſulas in eadem circunferentia cum tota terra contineri, & mare inter quaſcunque duas inſulas in tumorem, & tumulum quendam attolli. Vnde ſi deflueret, uniuerſam terram cooperiret, etiam illam, que apud Antipodes eſt, unà cum omnibus inſulis. Verum haec reſponſio abſurda eſt. Primum, quia ſi ita eſſet, non haberet tota aqua unicum centrum, ſed quilibet tumulus aque inter duas inſulas ſuum proprium, quod eſt contra communem omnium ſententiam, & temere uidetur aſſertum* (our Eng. trans.).

> lowest parts. Thus, the earth is comparable to any wooden sphere perforated by a lot of hollows where the water can be received; in fact this globe has a weight so well calibrated that the centre of gravity and the centre of size match in it.[43]
>
> ...
>
> On the contrary, not only do the elements not minimally conserve this continuous decuple proportionality, but nothing else does [either], as Alexander Piccolomini correctly proves in the booklet on the quantity of the earth and the water; and the same thing is confirmed by Fernel in his *Cosmotheoria*.[44]

The question of the terraqueous globe was also covered, with an introduction of Euclidean geometry, by a contemporary of Clavius, Francesco Barozzi (1537–1604), author of the 1585 work *Cosmographia*.[45] Barozzi, a Venetian patrician, was born in Candia, a Venetian colony on the island of Crete, in 1537. As an adult he attended the University of Padua, where later on he also lectured. His cultural activity can be set in the Renaissance movement dedicated to a profound, critical study of the ancient science.

The *Cosmographia*, the work consists of two parts: the 350 pages which constitute the four books of the *Cosmographia* are preceded by 64 pages of densely set type with the title *Errores Ioannis de Sacrobosco et eius expositorum et sectatorum*, with the addition of another 53 pages with the title *Communia Mathematica Principia*. In this singular writing Barozzi accuses Sacrobosco of having made 84 mistakes, which he identifies and discusses, pointing out the reasons for every error. According to him, these errors were repeated and even increased by the commentators, but he does not repeat them. To help novices understand his *Cosmographia*, he prefaced it with an elementary exposition of the principles of mathematics: "*Sine quibus fieri non potest, ut a Tironibus, in Euclidis, ac Teodosij Elementa minime versatis, ea quae docturi sumus perfecte percipiantur*".

This singular criticism of Sacrobosco was not appreciated by Clavius, who went so far as to wrote him in a letter that those "trifles of Sacrobosco" ruined the

---

[43]Christopher Clavius: *In sphaeram…*, p. 126: *Ad hanc obiectionem dicendum est, eam ex falsa hypothesis procedere; putat enim ex una tantum parte esse terram, et ex apposita totum mare, quod falsum est Navigationibus enim huius nostrae tempestatis tam sub polis, quam sub Aequinoctiali circulo, tam in oriente, quam in occidente, et denique in toto orbe reperta sunt vel continentia, vel insulae, vel peninsulae, ita ut per totum orbem fere permixtae sint terrae et aqua. Est enim mare innumeris pene insulis conspersum, adeo ut plus terrae extra mare appareat, quam aquis sit contectum, ut egregie probat Alexander Piccolomineus in libello de quantitate terrae et aquae. Unde dicimus hunc globum, quem confici asserimus ex terra et aqua, ita esse comparatum, ut terra circumquaque emineat, aqua vero in partibus humilioribus desidat. Referet itaque terra globi cuiusdam lignei speciem, in qua plurimae sunt concavitates, in quibus aqua possit recipi; nam sic aequalitate ponderium ita est hic globus collibratus, ut idem habeat centrum gravitatis cum centro magnitudinis* (our Eng. trans.).

[44]Christopher Clavius: *In sphaeram…*, p. 130: *Immo non solum elementa hanc proportionem continuam decuplam minime observant, sed nec aliam continuam, ut recte probat Alexander Piccolomineus in opusculo de quantitate terrae et aquae; idem, que confirmat FerneliusAmbianas in sua* Cosmotheoria (our Eng. trans.).

[45]Francisco Barocio, Iacobi filio, Patritio Veneto autore (Francesco Barozzi), *Cosmographia in quatuor libros distributa* (Venice: Gratiosus Perchacinus excudebat, 1585).

reputation of the work: "…your *Cosmographia* would enjoy a better reputation without these bagatelles of Sacrobosco". In Clavius' opinion there were no errors but only "improprieties" and "transpositions of the subjects", that is, only formal inaccuracies.[46] Of course we are not interested in this matter, and we have only quoted it as a curiosity. Now let us pass to consider the heart of the work.

Barozzi begins by lengthening the title, which becomes *De Sphaera Mundi sive Cosmographia*, and structures it as a complete description of the universe. Following Aristotle, he divides the machine of the universe into two parts: the celestial and the terrestrial (that is, made of elements). The celestial part consists of the ten orbs and is unalterable; the terrestrial part, which is continuously subjected to alteration,

> … is divided into four bodies, according to the four convenient constitutions of the four primary qualities, that is Earth cold and dry, Water cold and humid, Air humid and hot, Fire hot and dry.
>
> Of which the earth is as the centre of the world in the midst of all, encircled and covered in many of its parts by the water, remaining the most of it uncovered, the quantity of water not being sufficient to cover it completely. Thus it was established by the omnipotent and glorious God for preserving the life of living beings.
>
> Thus the earth and the water together construct a perfectly spherical machine, that is a globe, which is surrounded all around by the air and in the same way the air is spherically surrounded by the fire.[47]

Forty pages later he writes:

> The quantity of water alone cannot be known, since it cannot be measured alone, if not together with earth with which (as we already said) it constructs a perfectly spherical globe. But it has been well demonstrated by a great number of most experienced people that the water is much less than the earth both in the surface and in the body: and it is because of this that it cannot in any way cover the earth completely, but the surface of the earth not covered by waters is larger than the covered one.[48]

---

[46]The letter, dated 29 January 1586, can be seen in U. Baldini, P. Napolitani (eds): *Christoph Clavius Correspondenza*, vol. I (Università di Pisa, Dip. di Matematica, 1992), pp. 65–66 (lettera 26) (our Eng. trans.).

[47]Francesco Barozzi, *Cosmographia*, I, II (p. 3 of the 1598 edition): *In quatuor corpora, iuxta quatuor primarium qualitatum convenientes complexiones, dividitur; videlicet Terra frigida et sicca: Aquam frigidam et humidam: Aerem humidum et calidum: et Ignem calidum, et siccum. Quorum terra est tanquam centrum mundi in medio omnium sita, circumdata, et cooperta iuxta multas suas partes ab aqua, relicta maiori eius parte detecta, cum non sit aquae tanta quantitas quae omnino eam cooperire possit. Quod ita a Deo omnipotenti,et glorioso constitutum est ad vitam Animantium conservandam. Terra autem simul cum aqua machina una perfecte sphaerica, sive globus unum conformant quippe qui aëre undequaqua; et consimiliter aër ab igne spherice circumdatur* (our Eng. trans.).

[48]Francesco Barozzi, *Cosmographia*, I, II (p. 43 of the 1598 edition): *Quantitas vero solius sciri minime potest, quum nullo pacto sola potest mensurari nisi modo iam dicto simul cum terra, quae una cum aqua (ut iam diximus) globum unum perfecte sphaericum formant. Demonstrastum autem est a plerique peritissimis Recentioribus aquam esse multo minorem quo ad superficiem, et quo ad corpus ipsa terra: propterea que; totam ea cooperire nequaquam potest, sed superficies eius aquis detecta maior est ea, quae aquis cooperitur* (our Eng. trans.).

As an aside we note that among the experienced people mentioned by Barozzi are Alessandro Piccolomini and Christopher Clavius. To these, who believed in the sphericity of the Earth, we can certainly add others who, still in the sixteenth century, continued to make use of the model of the homocentric spheres, including, for instance, Alessandro Achillini (1463–1512),[49] Girolamo Fracastoro (1478–1553),[50] Giovan Battista Amico (1511–1538).[51]

4.4 Willebrord Snell and the First Triangulation

Ninety years after the measurement of the terrestrial meridian by Jean Fernel, in Holland Willebrord van Royen Snell (Latinized as Snellius) published, in his book *Eratosthenes batavus*,[52] the results of the measurements personally executed by himself starting from 1614, by using a new method: triangulation. As we know, this method is still in use today (by means of an accurate instrument: the theodolite) and Snell, together with Eratosthenes, can be considered the founder of geodesy.

Centuries earlier the ancient Greeks had already used a trigonometric method for measuring distances between points which could not be connected directly (recall the episode of the tunnelling of Samos[53]).

In triangulation, the trigonometric method of measurement of distances is applied repeatedly. By starting from a baseline effectively measured on the ground by means of a measuring rod, through measurements of angles one obtains the measure of a second baseline much longer than the first; from this one obtains a third baseline and so on, until the measurement of the desired distance is obtained. See, for the sake of illustration, Fig. 4.3, where AB is the baseline of known length and C a point afar but in full view.

Then, by means of a quadrant angles ABC and BAC are measurable, and, using trigonometry, AC and BC are also measurable. These sides, in their turn, can be used as baselines from which to sight C_1 and C_2 and get new baselines AC_1 and BC_2. Of course, as reference points for the application of the method one takes points that are increasingly far apart but easily sighted, such as the steeple of a church or a mountain top, thus making it possible, by sighting them with a quadrant, to measure the angles.

[49]With the work *De Orbibus* (1498).

[50]With the work *Homocentrica* (1536).

[51]With the work *De Motibus Corporum Coelestium iuxta principia peripatetica sine eccentricis et epiciclis* (1536). See also N. Swerdlow: "Aristotelian Planetary Theory in the Renaissance: Giovanni Battista Amico's Homocentric Spheres", *Journal for the History of Astronomy* 3 (1972): 36–48.

[52]Willebrordo Snellio [Willebrord Snell], *Eratosthenes batavus, de terrae ambitus vera quantitate* (Leiden: George Abrahamsz van Maarssen for Jodocus Colster, 1617).

[53]For an account of this, see G. Pólya *Mathematical Methods in Science*, op. cit., pp. 3–6.

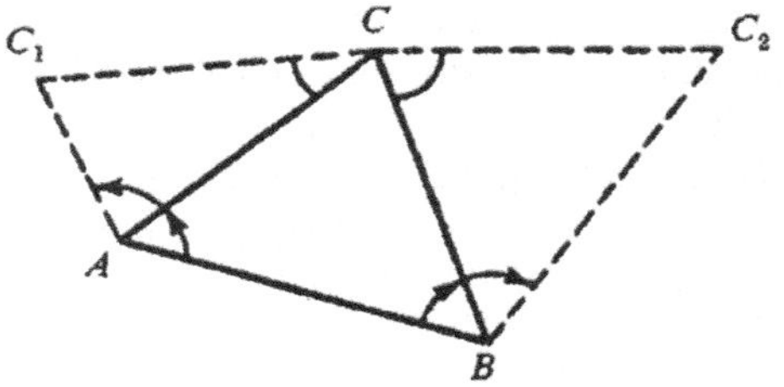

Fig. 4.3 The procedure for calculating distances by triangulation

Before Snell, the use of the method of triangulation was suggested by Gemma Frisius (1508–1555), who in 1533 published an enlarged edition of Peter Apian's *Cosmographia*. There he included his work *Libellus de locorum describendorum ratio*, which in particular contains the first proposal to use triangulation as a method of accurately locating places. Following Gemma's teaching, Gerardus Mercator made the survey of Flanders.

Let us now pass to Snell and his work. He was born in Leiden in 1580, son of a professor of mathematics in the local university. In the same university he became full professor in 1615. Before that time, he had travelled to various European countries, visiting several mathematicians and astronomers, including Tycho Brahe, Michael Maestlin and Johannes Kepler. In Prague with Brahe he spent some time assisting him in making observations. This experience came to an end in October 1601 when Brahe died. Snell continued visiting mathematicians in various German towns, returning to Leiden in 1602. Successively he went in Paris to study law. Afterwards, however, he gave up law and spent the rest of his life in Leiden, where he died in 1626. Today he is known to anyone who studies physics as the author of the law of refraction (Snell's law), discovered in 1621 but only made public in 1703 when Christiaan Huygens published it in his work *Dioptrica*.

Our interest is devoted to *Eratosthenes batavus* and to Snell's activity in the measurement of the terrestrial meridian. As we have said, Snell started his measurements in 1614. By means of a measuring rod of length 3.768 m (1 Rhineland rod) he measured around Leiden a baseline of 328 m. Sighting from the extremities of this baseline by means of a quadrant with a radius of 60 cm, he obtained the measure of other three baselines all in the vicinities of Leiden and Zoeterwoude (using as reference points the tower of the Cathedral of Leiden and the church steeple in the village of Zoeterwoude). By repeating the procedure over and over, he first obtained the distance between Leiden and the Hague (15,800 m) and then, with a series of triangulations, he succeeded in measuring the distance between Alkmaar (in the north of the Netherlands) and Bergen Op Zoom (in the south): 1,307,201 km. Finally, Snell had to find the geographical latitudes of Alkmaar, Bergen Op Zoom and Leiden, which he did by measuring the height of the Pole Star.[54]

[54]The reader can find a recent exposition of the details of all these operations in Liesbeth de Wreede: "Willebrord Snellius (1580–1626) a Humanist Reshaping the Mathematical Sciences",

Following this, for orienting his triangle network, he also had to solve a geometrical problem which is still present today in the activity of surveying: the three-point resection problem. This consists of finding the location of an observer by measuring the angles subtended by lines of sight from this observer to three known stations. Snell had to determine the distance of his house (in Leiden) to three points in Leiden, the mutual positions of which were known. He solved the problem (today his method is referred as the Pothonot-Snellius method) and proudly announced his solution:

> I have invented an elegant theorem for that problem, which can have a widespread application in our country from now on, because the distances between so many illustrious places have been registered with such precision.[55]

The entire triangulation by Snell consisted of 33 triangles which were used, essentially, in the way still customary today. The final result for the length of the meridian was 38,639 km (about 3.65% less than the modern value). These were, in summary, the measurements effected over the course of about three years.[56] Successively, Snell, together with his collaborators, effected further measurements, having found some mistakes, started to work at an improved version of *Eratosthenes batavus*, but this, however, was never published. In fact, Snell died before having brought his work of revision to its conclusion. The drawing up of the new measures was only completed a century later, and it turned out that Snell's revision gave a length of 40,370 km for the terrestrial circumference, indeed an admirable result if one considers the instruments at his disposal.

Let us now look at the *Eratosthenes batavus* which, from its very title, reveals Snell's idea of writing an important work destined to play a fundamental role in the history of the measurements of the Earth. In fact, he styles himself as the Dutch Eratosthenes. The work consists of two books: the first (pp. 1–116) reviews both the theoretical speculations and the measurements carried out up to the last one (that of Fernel); the second (pp. 117–264), after a comparison of the various units of measurements used at the time, explains the measurements performed and the methods used. Like all the printed works of that period, the work begins with a dedicatory letter, in this case addressed to the States General, where he claims the importance of knowing the measure, as much as possible accurate, of the size of the Earth. He proudly proclaims:

Ph.D. thesis, Utrecht University, 2007 (available at: https://dspace.library.uu.nl/handle/1874/22992).

[55]Willebrord Snell, *Eratosthenes batavus*, p. 199: *Et ad eam rem theorema scitum excogitavi, cuius usus iam in patria nostra deinceps permagnum esse possit, cum tot illustrium locorum intervalla tam accurate sint consignata* (Eng. transl. Liesbeth de Wreede; see that thesis for all the details).

[56]The reader can find all this in de Wreede, "Willebrord Snellius", op. cit., Chap. 3, which includes an exhaustive bibliography (pp. 329–365).

> I have tackled a problem the solution of which has always been desired by everyone, which has been tried very often, and which has also been made famous by the endeavours of great men. … I present here an accurate assessment of the size of the globe.[57]

We observe that, in the review that Snell has done of the measurements performed in the past, he has not left out those performed by the Arabs (Chap. XX, *De Arabum Geodesia*), of which he has found information in the work of an Arabian writer, Abelfedeas (1322). Before setting forth his researches, he wishes to survey the state of the art on the problem scrupulously. He shows himself to be very scrupulous indeed, comparing the units of measurement used in the various countries with the Rhenish foot (used in Holland), a task to which he devotes five chapters (pp 121–156) starting from the first (*Pedis Rhynlandici ad aliarum gentium pedes comparatio*). He notices that his result can be used by other people only if the way of translating it in the units of measurement of the various countries is given.

We have already mentioned that he found a rigorous geometric solution for the three point resection problem showing evidence of his mathematical skill but, for what regards the spherical triangles, he determines not to apply the spherical trigonometry. In fact, he concludes that, since the measured distances are so small with respect to the radius of the Earth, "...*errorem nullum inducere potuerit*".[58]

Of course, since the second book is also a kind of diary of the measurements, he does not avoid mentioning the difficulties of the undertaking:

> "if the general profit, and the illustrious energy invested on this topic in so many centuries before me, had not spurred me on and forced me to take my pen in hand again, and to raise my body and the sharpness of my eyes to the peaks of towers … What I report here is hardly the hundredth part of the exertion, trouble and expenses which I have endured.[59]

Summing up, we could define Snell as an "applied mathematician", that is, a mathematician who passes from researches in pure mathematics (for instance the search of the approximate value of π), to the laws of optics, to the edition of Apollonius' work, and finally to measurements on the ground by developing the mathematics necessary for obtaining increasingly accurate measures. In addition, his measurements are not the researches of a scientist locked in his ivory tower (in this case, the University of Leiden), since his triangulations are supported by the public administration (hence the dedicatory letter to States General). Whereas the measurements of Fernel ("*medicus celeberrimus*" as Snell calls him) are—so to say

---

[57]Willebrord Snell, *Eratosthenes batavus*, pp.)?(iii^{v}—)?(iiii^{r}: *Rem aggressi sumus ab omnibus semper desideratam, saepius tentatam, et magnorum quoque virorum industria nobilitatam … Orbis terrae quantitatem accurate definitam hic exhibeo* (Eng. transl. Liesbeth de Wreede).

[58]Willebrord Snell, *Eratosthenes batavus*, p. 198.

[59]Willebrord Snell, *Eratosthenes batavus*, p. 171: *nisi publica utilitas, et tot iam seculis fatigata tam nobilis cura stimulos mihi addidisset, et rursum calamum in manum, corpus et oculorum aciem in turrium fastigia attollere coegisset … Haec enim ipsa quae hic affero vix centesima pars sunt laboris, molestiae impensarum quas exantlavimus* (Eng. transl. Liesbeth de Wreede).

—the amateur exploit of one who practises another profession, in the case of Snell we are in the ambit of what we would define a public work: science at the service of public administration.

If on the one hand Snell's measurements integrate astronomical knowledge, on the other hand they appertain to the public interest. In fact, he surveyed a large part of Holland and the surrounding provinces, enabling the States General to accurately record their home country. A further important consideration: we are now in the period in which the universities are free of ecclesiastical wardship and scientific research is "secular research" (at least in most cases).

4.5 Jean Picard and the Beginning of the French Triangulations

Around fifty years after the appearance of *Eratosthenes batavus*, in France (under the reign of Louis XIV) Jean Baptiste Colbert (1619–1683), Finance Minister, was instructed by the king to found an academy of scientists, guaranteeing them a wage and funds for their researches. Thus was born, in 1666, the Académie Royale des Sciences (Royal Academy of Sciences). The Dutch Christiaan Huygens was appointed director of the Académie. One of the principal problems of the day was the measurement of the Earth. Colbert was convinced that making of an accurate map of the France would have made it possible to improve the administration of the kingdom, and the measurement of the circumference of the Earth was the necessary basis for starting the work.

The task of organizing a campaign of measurements was assigned, in 1667, to abbot Jean Picard (1620–1682). He was born at La Flèche, a town on the banks of the Loire, and there he studied at the famous Jesuit College (where, before him, Renè Descartes had also studied). He left the Jesuit College at La Flèche around 1644 and went to Paris, where he attended the lectures of Pierre Gassendi at the Collége Royale. Afterwards he was Gassendi's assistant in various astronomical observations and in 1655 became professor of astronomy at the Collège Royale, following the death of Gassendi. In those years he also met Huygens and successively became a member of the Académie Royale des Sciences, just after its foundation. There are no traces of any publications by him before his entry in the Académie, where he devoted himself to working from that time on.

Picard greatly increased the accuracy of measurements of the Earth by perfecting the method of triangulation used by Snell. His results were published in a book with the title *Mesure de la Terre*,[60] which appeared in 1671. Regarding the title (in French) of this book, let us take the liberty of making a little "linguistic" digression. In the first decades of the seventeenth century, the position of Latin as the

[60]*Mesure de la Terre par M. l'Abbé Picard de l'Académie Royale des Sciences* (Paris: L'imprimerie Royale, 1671).

international language was already deeply shaken. That which had been the language of all learned men throughout the Middle Ages, which fit in with all the speculations of the Scholastics as with the precepts of the medical schools, and which the humanists had wanted to purify and restore to its Ciceronian perfection, had run its course and was no long capable of competing successfully against the national languages. As regards Italy, we have already mentioned the example of Alessandro Piccolomini, who in the middle of the sixteenth century had written "scientific" works in Italian. This trend was strengthened with Galileo's *Dialogue* in the third decade of the following century. Of course, the fight would last for some time to come (Newton wrote the *Principia* in Latin in the late seventeenth and early eighteenth centuries, and Gauss wrote various mathematical essays in Latin in the first part of the nineteenth century, which underlines the fact that English and German were still not "exportable" scientific languages), but in France, starting with Descartes (the *Discours de la Méthode* is of 1637), the situation stabilized rapidly and all the discussions on the shape and the size of the Earth which we shall now deal with were written in French.

Let us speak now about the work of Picard. *La mesure de la terre* (106 pages of small size with 13 articles) is much shorter than Snell's book. Picard devotes only the first article to summarizing the history of the preceding measurements, dwelling longer on the Arabs' measurements. He then mentions Fernel's measurements, pointing out polemically that Fernell does not divulge the name of the place to which a distance of 25 leagues to Paris is attributed. Of Snell, instead he says:

> *Snellius a tenu une méthode plus certaine, & semblable à celle qui se verra pratiquée dans la suite ... il a cherché par des voies géométriques les distances méridiennes....*[61]

At the end, he adds the consideration "*cette dernière mesure étoit communément suivie comme la plus exacte*", contrasting to it the measurement performed in Italy by Father Giovan Battista Riccioli (1598–1671) between Modena and Bologna, and postponing the criticisms of it until the end of the book.

He closes the first article by defending the necessity of new measurements both with regard to the longitudes, and "*particulièrement encore pour l'usage de la Navigation*" and further pointing out that until then no one had realized that the employment of the telescope would have carried a great advantage. Moreover, he maintains that one needs a much longer baseline than that used by Snell. Of course, for measuring the first baseline directly it is necessary to establish a unit of measurement and to that end he chooses the Paris toise (1 toise corresponds to 10 feet, 1 foot to 12 inches, 1 inch to 12 lines, and it was defined by an iron bar restored by Picard himself that was affixed to the façade of the Chatelet Palace in Paris). In order to define it in a way which is repeatable everywhere, he says "*nous l'attacherons à un Original, lequel étant tiré de la Nature même doit être invariable & universel*".[62]

[61]Jean Picard, *La mesure de la Terre*, op. cit. p. 8.

[62]Jean Picard, *La mesure de la Terre*, op. cit. p. 15.

What is this "*Original*"? It is the length of a seconds pendulum, that is one which has the period of one second (defined through the mean motion of the Sun). As we know, the period T of a simple pendulum of length l (in the approximation of small oscillations) is given by $T = 2\pi\sqrt{\frac{l}{g}}$, where g is the gravitational acceleration, and then $l = \frac{gT^2}{4\pi^2}$. If the period is fixed, the length is directly proportional to g. If g remains constant everywhere on the terraqueous globe, then also the length of the seconds pendulum is a universal constant and therefore it is possible to translate the Paris toise defined through it into any other unit of measurement of different countries (the length of the pendulum with period equal to 1 s measured in terms of the Châtelet toise, was 36 inches 8½ lines). If, on the contrary, g is not a constant, the weight of a body of mass m, $p = mg$, is in its turn proportional to the length of a seconds pendulum.

At this point (article IV), Picard reports that there are

> *quelques Expériences, d'où il semble que l'on pourroit conclure que le Pendules doivent être plus courts, à mesure qu'on avance vers l'Equateur, conformément à la conjecture qui avoit esté déja proposée dans l'Assemblée, que suppose le mouvement de la Terre, les Poids devroient descendre avec moins de force sous l'Equateur que sous les Poles* [*mais*] *nous ne sommes pas suffisamment informés de la justesse de ces Expériences, pour en conclure quelque chose.*[63]

The hypothesis that a body is heavier at the Pole than at the Equator (the centrifugal force, which opposes to the gravitational attraction, is proportional to the radius) is here published for the first time and then, for the moment, put aside. In any case, Picard says,

> *S'il se trouvoit par Expérience que les Pendules fussent de différente longueur en différents lieux, la supposition que nous avons faite touchant la mesure universelle tirée des Pendules, ne pourroit subsister; mais cela n'empescheroit pas que dans chaque lieu il n'y eût une mesure perpétuelle & invariable.*[64]

After this, he continues (starting from the fifth article) by describing first of all the instrument used in the measurements of the Earth, since, as he says it "*a quelque chose de particulier*". It was a quadrant with a radius of 38 inches with two telescopes, one fixed and the other revolving equipped with a crossing of threads in the focus of the objective and of the ocular. The baseline to be measured on the ground was taken on a sufficiently straight stretch of road between Villejuive and Juvisy, south of Paris.

This length was measured by means of two planks of wood (two Paris toises in length = 390 cm) carried forward alternatively following a stretched rope. The measurements of the angles, necessary for calculating the sides of the various triangles, were performed with great accuracy. As an example, the longest side (that which joined Malvoisine to Mareuil) was determined both directly (62,199 km) and

[63] Jean Picard, *La mesure de la Terre*, op. cit. pp. 19–20.

[64] Jean Picard, *La mesure de la Terre*, op. cit. pp. 19–20.

Fig. 4.4 Illustration of night-time sighting the targets for measuring

indirectly (62,192 km) through the combination of other triangles: (only 7 m of difference and an uncertainity of 0.1‰ (the same obtained in the measurement of the baseline on the ground).

It is interesting to note how the direct observation was executed. The targets were sighted at night, by lighting great beacons in their proximity. This way of carrying on is illustrated in the book with an engraving at p. 3 above the title (Fig. 4.4).

The distance that Picard intended to measure was that between the parallels of Malvoisine (south of Paris) and Sourdon (north). The aim was achieved with 13 triangles as regards the measurements on the ground. The difference of latitude (i.e., the amplitude of this arc) was determined by measuring, at both Sourdon and Malvoisine, the zenithal distance of star δ of Cassiopeia.

The result for 1° of meridian was 111,212 km, from which one obtains 40,033 km for the meridian circumference of the Earth (with an uncertainty of 1/10,000). Obviously, it is beyond our aim to report a detailed account of Picard's measurements. We shall limit ourselves to commenting on his criticisms of his predecessors. At the end of the book (the thirteenth article) he begins with Fernel and then continues with Snell and Riccioli. As regards Fernel, he cannot hide his disappointment for the good result obtained by means of raw procedures:

> *Il y a sans doute de quoi s'étonner que par une maniere aussi grossière que la sienne, il ait approché si près de la mesure que tant d'Obsservations nous ont fait conclure.*[65]

He then concludes that the errors in the astronomical observations were counterbalanced by those in the length of the road. As regards Snell, he repeats that the fundamental error was that of having taken an initial baseline that was too small, leading to a great number of triangles and angles that were too small. Finally, he

[65]Jean Picard, *La mesure de la Terre*, op. cit. p. 99.

devotes much more space to discussing and criticizing the measurements of Father Riccioli, maybe because at that time Riccioli was considered an authoritative scientist.

4.6 Conclusion

The two centuries (from the middle of the fifteenth through the first half of the seventeenth) which we have covered in this chapter very rapidly, but establishing the points we consider fundamental, carry us from what can be considered the "experimental proof" of the sphericity of the Earth (Magellan etc.) to the accurate measure of its size (in the hypothesis of a perfect sphere). In the first case, there is simply an experience "*in corpore vili*", since the roundness of the world is really covered. In the second, the progress of technology (optical instruments) makes it possible to obtain results of indisputable quality. With Picard, one can say that we have arrived at the end of what can be called the spherical era of geodesy that began with Eratosthenes. As we shall see, the ellipsoidal era will start with the experiences of Jean Richer at Cayenne and with the theory of gravitational attraction by Newton, who for this, inter alia, will use Picard's results.

Suggested Readings

Burckhardt, J. (1860). *The civilization of the renaissance in Italy* (1st German edn). 1860, various reprints of the English translation.

Cliff, N. (2011). *Holy war: How vasco da gama's epic voyages turned the tide in a centuries-old clash of civilizations*. Harper Collins.

Columbus, C. (1992). *The four voyages* (J. M. Cohen, Trans.). Penguins Classics, rpt.

Garin, E. (1969). *Science and civil life in the Italian renaissance* (1st edn). Anchor Books.

Garin, E. (1983). *Astrology in the renaissance: The zodiac of life*. Viking Press.

Wootton, D. (2016). *The invention of science: A new history of the scientific revolution*. Harper Collins.

Chapter 5
The Shape and the Size of the Earth in the Eighteenth Century

5.1 The Observations of Richer at Cayenne and Their Consequences

The last decades of the seventeenth century were particularly important for the astronomical observations and their correlated theories: recall that a few years before the foundation of the Académie Royale des Sciences in Paris, the Royal Society of London was also founded (1662), and that the two competed in obtaining successes in the scientific field. After Picard's undertaking, which resulted in what at that time was the most reliable measurement of the terrestrial meridian, the Académie undertook to carry out several projects of scientific research also in collaboration with the great Observatory that had been constructed outside Paris and to whose direction had been appointed in 1671 the famous Italian astronomer Gian Domenico Cassini (who, later on, was naturalized French and became Jean Dominique Cassini).

We are not interested here in talking about the various projects and estimated astronomical observations, but we wish to focus only on the observations done by Jean Richer on the island of Cayenne in French Guyana in 1762.[1] We do not have biographical data about Richer, apart from his activity in the Académie Royale des Sciences, but it is conjectured that he was born in France in 1630. He became a member of the Académie in 1666 with the title of *élève astronome* and by 1670 he had also been given the title *mathématicien*.[2] He was entrusted with various missions by the Académie but the one which is universally mentioned for the result which will be fundamental in the discussions about the shape of the Earth is the one carried out in Cayenne.

---

[1]Richer gave a detailed account of his researches in the book *Observations Astronomiques et Physiques faites en l'Isle de Caienne* (Paris: Imprimerie Royale, 1679).
[2]See the article by J. J. O'Connor and E. F. Robertson in the MacTutor History of Mathematics archive, http://www-history.mcs.st-and.ac.uk/Biographies/Richer.html.

D. Boccaletti, *The Shape and Size of the Earth*,
https://doi.org/10.1007/978-3-319-90593-8_5

The mission to Cayenne in 1672–73 took place after the Richer's homecoming from another important mission in Canada in 1670,[3] which had led to some disagreements with Huygens regarding the clocks which Huygens himself had constructed for the determination of longitude at sea. Richer was to lead the expedition, assisted by M. Meurisse, and discussed the final details of the research to be undertaken with Jean Dominique Cassini in the Paris Observatory.

There were three questions with which the expedition was concerned: the movements of the Sun and the planets, refraction and parallax.[4] A determination of the solar parallax, a fundamental astronomical constant, would open the way to the long-sought knowledge of the actual dimensions of the solar system. Certainly, at the outset, nobody had expected an investigation of the length of a seconds pendulum. Instead this will be the observation destined to go down in history.

The object of the expedition, as we have already said, was to execute astronomical observations that, in that period, it was become possible to execute with great accuracy. In fact, as James Olmsted says:

> Between 1650 and 1670 three great inventions, pendulum clock, filar micrometer, and the application of telescopes to the graduated circles of astronomical instruments in place of the conventional sights or *pinnules,* combined to effect a revolution in observational astronomy. An entirely new standard of accuracy and refinement became possible.[5]

But the pendulum clock was one of the fundamental instruments and Richer realized that there was something that did not make sense. He devoted the Article I of the tenth chapter of his work (entitled "*De la longueur du pendule à seconds de temps*") to this fact:

> One of the most important observations I have made is that of the length of the seconds pendulum, which has been found shorter in Cayenne than at Paris: for the same measure which had been marked in that place – there on a rod of iron according to the length found to be necessary to make a pendulum of a second of time, was transported to France and compared with the Paris measurement. The difference between them was found to be $1^1/_4$ lines, by which the Cayenne measurement falls short of the Paris measurement, which is 3 feet, $8^3/_5$ lines. This observation was repeated during ten whole months, when no week passed without its being carefully performed several times. The vibrations of the simple pendulum which was used were very short and remained quite perceptible up to 52 min, and were compared with those of an extremely good clock whose vibrations indicated seconds.[6]

---

[3]On this see J. W. Olmsted: "The voyage of Jean Richer to Acadia in 1670: A study in the relations of Science and Navigation under Colbert" *Proceedings of the American Philosophical Society* 104 (1960): 612–634.

[4]See J. W. Olmsted: "The Scientific Expedition of Jean Richer to Cayenne (1672-1673)", *Isis* 34 (1942): 117–128.

[5]J. W. Olmsted: "The Scientific Expedition", op. cit., p. 123.

[6]J. Richer, *Observations astronomiques et Physique*, op. cit. p. 66: *L'une des plus considerables Observations que j'ay faites, est celle de la longueur du pendule à secondes de temps, laquelle s'est trouvée plus courte en Caïenne qu'à Paris: car la mesme mesure qui avoit esté marquée en ce lieu – là sur une verge de fer suivant la longueur qui s'estoit trouvée necessaire pour faire un pendule à secondes de temps, ayant esté apportée en France, & comparée avec celle de Paris, leur*

The subject in itself was of great interest and was also studied on the other side of the English Channel, especially by Newton who already had used Picard's measure of the meridian. To it he devoted (already in the first, 1687 edition of the *Principia*[7]) Proposition 20 (pp. 424–426). In the third edition of 1726,[8] he even reported Richer's measure by paraphrasing his book:

> Now some astronomers, sent to distant regions to make astronomical observations, have observed that their pendulum clocks went more slowly near the equator than in our regions. And indeed M. Richer first observed this in the year 1672 on the island of Cayenne. For while he was observing the transit of the fixed stars across the meridian in the month of August, he found that his clock was going more slowly than in its proper proportion to the mean motion of the sun, the difference being $2^{m}28^{s}$ every day. Then by constructing a simple pendulum that would oscillate in seconds as measured by the best clock, he noted the length of the simple pendulum, and he did this frequently, every week for ten months. Then, when he had returned to France, he compared the length of this pendulum with the length of a seconds pendulum at Paris (which was 3 Paris feet and $8^3/_5$ lines long) and found that it was shorter than the Paris pendulum, the difference being $1^1/_4$ lines.[9]

As we have already mentioned, the length of the seconds pendulum is directly correlated to gravitational acceleration and then to the weights of the bodies in the various places of the Earth (see §4.5).

In the *Principia*, Proposition 20 ("To find and compare with one another the weights of bodies in different regions of our earth", from which we have taken the excerpt above), followed Propositions 18 ("The axes of the planets are smaller than the diameters that are drawn perpendicularly to those axes") and 19 ("To find the proportion of a planet's axis to the diameters perpendicular to that axis").

We believe expedient to quote the short Proposition 18:

> If it were not for the daily circular motion of the planets, then, because the gravity of their parts is equal on all sides, they would have to assume a spherical figure. Because of that circular motion it comes about that those parts, by receding from the axis, endeavor to ascend in the region of the equator. And therefore if the matter is fluid, it will increase the diameters at the equator by ascending, and will decrease the axis at the poles by descending. Thus the diameter of Jupiter is found by astronomical observations to be shorter between the poles than from east to west.[10] By the same argument, if our earth were not a little

difference a esté trouvée d'une ligne & un quart, dont celle de Caïenne est moindre que celle de Paris, laquelle est de 3. pieds 8. lignes $^3/_5$. Cette Observation a esté reïterée pendant dix mois entiers, où il ne s'est point passé de semaine qu'elle n'ait esté faite plusieurs fois avec beaucoup de soin. Les vibrations du pendule simple dont on se servoit estoient fort petites, & duroient fort sensibles jusques à cinquante-deux minu-tes de temps, & ont esté comparées à celles d'une horloge tres-excellente, dont les vibrations marquoient les secondes de temps (our Eng. trans.)

[7] I. Newton, *Philosophiae Naturalis Principia Mathematica* (London: Jussu Societatis Regiae ac Typis Josephi Streater, 1687).

[8] I. Newton, *Philosophiae Naturalis Principia Mathematica Eq. Aur. Editio tertia aucta & emendata* (London: Apud Guil. & Joh. Innys, Regiae Societatis typographos, 1726).

[9] I. Newton, *The Principia*. A new Translation by I. Bernard Cohen and Anne Whitman (University of California Press, 1999), p. 829.

[10] In the first edition, Newton quoted the names of the astronomers: Cassini and Flamsteed.

higher around the equator than at the poles, the seas would subside at the poles and, by ascending in the region of the equator, would flood everything there.[11]

Thus it clearly emerges that the problem of the variation of the weights of the bodies with the latitude on the Earth is presented by Newton as a general characteristic of all planets caused by the rotation of the planets around their axes with consequent deformation of the shape (no longer spherical). To Newton (and to Huygens as well) the deformation consisted in a flattening at poles, but, as we shall see in what follows, this conclusion was not shared by all.

5.2 Newton's and Huygens' Theories of the Earth's Shape

Let us see now how Newton and, later on, Huygens held that the Earth's shape was that of a spheroid flattened at the poles. According to Newton, supported by Cassini's and Flamsteed's observations of Jupiter, all planets appear flattened because of the rotation around their own axis, otherwise gravity would keep them in a perfectly spherical shape. Newton started by assuming an Earth constituted from the beginning by a fluid sphere whose size he established by using the measure of the meridian obtained by Picard in 1669–70 from the measurement of the degree of latitude between Amiens and Malvoisine.[12]

Knowing, obviously, the Earth's sidereal rate of diurnal rotation, Newton calculated the centrifugal acceleration at the equator of this spherical Earth. He then calculated the gravity acceleration that there should have been at the equator of a perfectly spherical Earth.

Here is the sequence of Newton's calculations. First, he calculated the effective gravity acceleration at the latitude of Paris, taking advantage of the results of experiments on free falling bodies performed there with seconds pendulums. Knowing the latitude of Paris he was able to obtain the centrifugal acceleration at that latitude. The effective gravity acceleration plus the centrifugal acceleration gave the magnitude of the gravity acceleration that should have been at any point on the surface of a spherical Earth which attracts according to the universal inverse-square law. Newton then found the ratio of his calculated values of the two accelerations at the equator to be the fraction $^1/_{289}$.[13]

---

[11] I. Newton, *The Principia*, op. cit., p. 821.

[12] Actually, starting in the second edition of the *Principia* he made use of the more up-to-date measures obtained under the direction of Cassini in 1701. Since these measures were subsequently judged too large, in the third edition, he finally used a more accurate measure of the meridian between Dunkerque and Paris (which turned out to be different from Picard's measure only by one toise).

[13] Actually, in the first edition of the *Principia* (see 6, p. 424), Newton gives the ratio 1 : $289^4/_5$, however without showing the calculations. In the successive editions, $^1/_{289}$ agrees with the ratio obtained by Huygens, who had performed his calculations after having read the first edition of the *Principia*. We shall see further the theory and relevant calculations of Huygens. In any case the ratio $^1/_{289} \cong 0.3460$ is very close to the ratio that one can calculate today using the precise values of the magnitudes regarding the Earth. The difference begins from the third decimal figure.

Even if the method followed by Newton in the calculation of the ratio between the centrifugal and gravitational acceleration at the surface of the Earth at the equator appeared clear to all the scientists at that time, this was not the case for the successive calculations for determining the flattening of the terrestrial spheroid. As John Greenberg says, making reference to the passages of Book III of the *Principia* where Newton deals with the subject: "…the theory struck even the most reputable continental mathematicians of the day as incomprehensible". This was due to:

> …the gaps, the underived equations, the unproven assertions, the dependence upon corollaries to practically incomprehensible theorems in Book I of the *Principia* and the ambiguities of these corollaries, the conjectures without explanations of their bases, the inconsistencies, and so forth.[14]

Greenberg thoroughly studied Newton's calculation, pointing out the weak points and the unproven assertions, but concludes: "I explain why these apparent drawbacks are, historically considered, strengths of the Newton's theory of the Earth's shape, not weaknesses".[15]

We cannot go into details of this question and limit ourselves to outlining the method used by Newton. The problem is faced in Proposition 19 of Book III, cited above, and the solution is obtained by imagining a right-angled canal (of unit cross section) hollowed in the Earth rotating around its north-south axis (Fig. 5.1).

The two branches of the canal are full of water and, since the figure of the Earth, according to Newton, must be a figure of equilibrium, the water must also be in equilibrium, i.e., the two columns of water must have the same effective weight. We call a the semidiameter of the equator and q the polar semiaxis. If we want to calculate the effective weight of the two columns of water and to impose their equality, we must have $\rho a(1 - m)\rho q$, where m is the ratio $^1/_{289}$ and ρ the density of the water. One can add that, since the attraction on any particle at a distance $r < R_E$ from the centre is proportional to r (see *Principia*, Book I, Proposition 91, corollary 3) both the centrifugal acceleration and the gravitational acceleration are proportional to the distance from the center. Again making use of Prop. 91, Newton computes that the force of gravity at the pole Q is in the ratio 501:500 to the force of gravity at the equator A. Then he invokes "the rule of three". If a centrifugal force of $^4/_{505}$ will cause a difference in length of the two branches of $^1/_{100}$, then a centrifugal force of $^1/_{289}$ will produce a difference in length of 230:229. Hence Newton concludes that the ratio of the equatorial diameter of the Earth to the polar diameter is 230:229.[16]

The mathematical formalism of which Newton makes use consists in Apollonian geometry and the Euclidean theory of proportions, and because of this it is natural that it appears difficult to a reader used to the modern formalism. Instead, as

---

[14] J. L. Greenberg: "Isaac Newton and the Problem of the Earth's Shape", *Archive for History of Exact Sciences* 49 (1995): 371–391, quote on p. 371.

[15] Ibidem.

[16] See I. Newton, *The Principia*, 1999, op. cit., Guide: Chap. 10, p. 350.

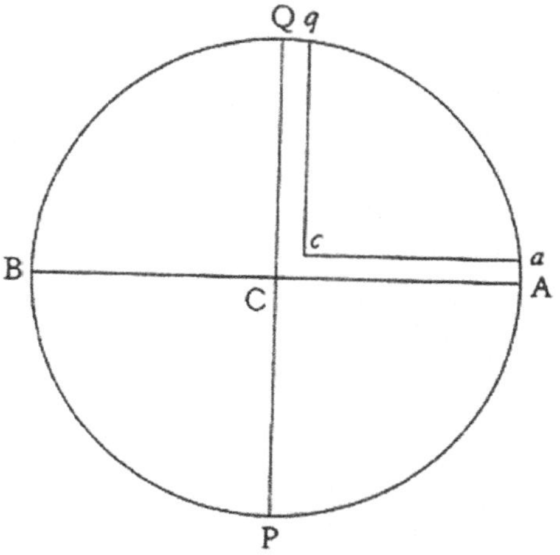

Fig. 5.1 Diagram of Newton's solution, obtained by imagining a right-angled canal (of unit cross section) hollowed in the Earth rotating around its north-south axis

Greenberg remarks, to his contemporaries what appeared difficult were the several assumptions that were neither justified nor demonstrated. However, in subsequent years most of them proved to be correct; Laplace, commenting on what he called "*the premier pas que l'on a fait dans la théorie mathématique de la figure de la Terre*", said:

> Malgré ces imperfections, ce premier pas doit paraitre immense, si l'on considère l'importance et la nouveauté des propositions que l'auteur établit sur les attractions des sphères et des sphéroides, et la difficulté de la matière.[17]

Let us see now what theory was advanced by the other great scientist of that time: Christiaan Huygens. First of all one must say that Huygens did not agree with Newton's theory of gravitation but drew inspiration instead from Decartes' vortex theory, even while not subscribing to it completely.[18] He explained his theory in a short work bearing the title *Discours de la Cause de la Pesanteur*, written in various times but only published in 1690 together with the more important *Traitè de la Lumière.*[19] The initial part of the work, which deals with the cause of the weight of bodies is described derogatorily by Isaac Todhunter:

> The theory of Huygens to account for Weight is ex-pounded on pages 129–144 of the work; we may say briefly that this theory is utterly worthless. Huygens assumes the existence of a very rare medium moving about the Earth with great velocity, not always in the same direction. This rare matter is surrounded by other bodies, and so prevented from escaping;

[17]*Traité de la Mécanique Cèleste par M. Marquis de Laplace* (Paris, 1825), vol. 5, p. 9.

[18]See: E. J. Aiton: *The Vortex Theory of Planetary Motions* (New York: American Elsevier Inc.: 1972).

[19]*Traité de la Lumière par C. H. D. Z.* [Christian Huygens de Zulichen] *Avec un Discours de la Cause de la Pesanteur* (Leiden, 1690). The *Discours* goes from p. 125 to p. 180.

> and it pushes towards the Earth any bodies which it meets. This vortex has passed away, as well as those similar but more famous delusions with which the name of Des Cartes is connected.[20]

Todhunter finally concludes: "The really valuable part of the *Discours* commences on page 145", where Huygens begins to talk about Richer's measures of 1762. The final part of the work (called *Addition*), which begins on p. 152, was written, as Huygens himself admitted, after having read the *Principia* and he too, for calculating the flattening of the Earth, makes use of the right-angled canal proposed by Newton. We find it expedient to start from the point where Huygens begins his talk about the centrifugal force and the importance it has in opposing gravitational acceleration:

> But I find, by my theory of circular motion, which agrees perfectly with experience, that given a body in circular motion, if one wants its effort for pulling away from the centre to equal exactly the effort of simple gravity, it is necessary that it performs each circuit in the same time in which a pendulum having the length of a semidiameter of this circle performs two swings. It is therefore necessary to see the time in which a pendulum having the length of the semidiameter of the Earth will perform two swings. This is easy, by the well-known property of pendulums, and for the length of that which beats seconds, which is of three foot and eight lines and ½, the measure of Paris. And I find that, for these two beats, a time of one hour and 24½ minutes should be necessary, by supposing, according to the accurate measure of Mr. Picard, the semidiameter of the Earth to be 19615800 foot. The velocity therefore of the fluid matter with respect to the surface of the Earth, must be equal to that of a body which should perform the circuit of the Earth in this time of one hour, 24½ minutes. Which velocity is almost exactly 17 times greater than that of a point on the equator which performs the same circuit, with respect to the fixed stars, as it must be taken here, in 23 h, 56 min.[21]

Let us explain Huygens' calculation. He starts from Picard's measure of the terrestrial meridian which we already know and which allows one to obtain as the radius of a spherical Earth 1961.5800 Paris feet (in modern units of measurement: $R_E = 6371.9965$ km, with a Paris foot = 32.484 cm). At this point, he asserts that if one wants that a body, which moves with uniform circular motion, be subjected to a

---

[20] I. Todhunter: *A History of the Mathematical Theories of Attraction*, op. cit., p. 29.

[21] C. Huygens, *Discours de la Cause de la Pesanteur*, op. cit. p. 143: *Mais je trouve par ma Theorie du mouvement Circulaire, qui s'accorde parfaitement avec l'experience, qu'un corps tournant en cercle, si on veut que fon effort à s'éloigner du centre, égale justement l'effort de fa simple pesanteur, il faut qu'il faffe chaque tour en autant de temps, qu'un pendule, de la longueur du demi diametre de ce cercle, en emploie à faire deux allées. Il faut donc voir en combien de temps un pendule, de la longueur du demidiametre de la Terre, seroit ces deux allées. Ce qui est aisé par la proprieté connue des pendules, & par la longueur de celuy qui bat les Secondes, qui est de 3 pieds 8½ lignes, mesure de Paris. Et je trouve qu'il faudroit pour ces deux vibrations 1 heure 24½ minutes; en supposant, suivant l'exacte dimension de Mr. Picard, le demidiametre de la Terre de 19615800 pieds de la mesme mesure. La vitesse donc de la matiere fluide, à l'endroit de la surface de la Terre, doit estre égale à celle d'un corps qui seroit le tour de la Terre dans ce temps de 1heure, 24½ minutes. Laquelle vitesse est, à fort peu pres, 17 fois plus grande que celle d'un point sous l'Equateur, qui fait le mesme tour, à l'égard des Etoiles fixes, comme on doit le prendre icy, en 23 heures, 56 min* (our Eng. trans.).

centrifugal force equal to its weight, it must make a complete circle in the same time that it takes a pendulum with the length of the radius of the circle to perform a complete oscillation. A body of mass m on the surface of the Earth, and thus of weight $p = mg$, following the circular motion of the Earth, is subjected to a centrifugal force mw^2R_E, where w is the angular velocity of the Earth's rotation and R_E is its radius. Its period of rotation is $T = \frac{2\pi}{w}$ and, if we set $p = mg = mw^2R_E$, we will have $T = 2\pi\sqrt{\frac{R_E}{g}}$. Comparing this with the known formula of the pendulum of length l, having the same period, we have $l = R_E$. Huygens makes use of this last relation for calculating the period of the pendulum having the length of the semidiameter of the Earth, by establishing a proportion with the known data of the Paris seconds pendulum. He obtains 1 h 24½ min = 5070 s. Comparing it with the diurnal period with respect to fixed stars (23 h 56 min = 86,160 s), he obtains $\frac{86,160}{5070} \sim 1699$, which he rounds up to 17. What we are interested in is the ratio between the centrifugal acceleration at the equator and the gravitational acceleration (for a spherical Earth). Therefore this ratio will be the inverse square of 17; i.e., $\frac{1}{289}$.

We have already remarked that one obtains almost exactly the same result, making use of the modern data and of the Newtonian gravitation theory. Finally we report that, in the *Addition*, Huygens follows Newton's method of considering the equilibrium of the two branches of the right-angled canal shown in Fig. 5.1 for calculating the ratio of the axes of the spheroid.

However, since his theory of gravity for the interior of the Earth was different from that of Newton, he found a different ratio: $\frac{577}{578}$.

We must also point out another fundamental result of Huygens. Again in the brief *Discours de la Cause de la Pesanteur*, he establishes an important principle which has in fact ever since generally been called by his name. He says the surface of the sea is "such that at every point the direction of the plumb line is perpendicular to the surface.[22] As Todhunter says, it can be stated more generally thus: "the direction of the resultant force at any point of the free surface of a fluid in equilibrium must be normal to the surface at that point".[23] In conclusion, we can maintain that, at the end of the seventeenth century and at the beginning of the eighteenth, the two most important European scientists agreed in thinking that the Earth, because of its diurnal rotation, had to have the shape of a flattened spheroid.

However, it did not take long for an opposing idea to arrive, especially in France: the Earth has an elongated shape. The principal promoter of this thesis was Jean Dominique Cassini. A controversy was born and blazed up in the first four decades of the eighteenth century.

---

[22]C. Huygens, *Discours de la Cause de la Pesanteur*, op. cit. p. 152: *Partant la surface de la mer est telle, qu'en tout lieu le fil suspendu est perpendiculaire* (our Eng. trans.).

[23]I. Todhunter, *A History of the Mathematical Theories of Attraction*, op. cit., p. 30.

5.3 The Peru and Lapland Expeditions

We should be tempted to say that, with Newton's theory, it is finally the turn of mathematical physics to speak: there is a theory which not only interprets the outcomes of the astronomical observations (the flattening of Jupiter), but by starting from a measurement (that of Picard), formulates a model which, in turn, can be compared with experimental data. Unfortunately in the historical reality things were not as easy as this.

The Académie promoted a campaign of measurements to determine the length of the degree of meridian which, for several years (starting in 1683), had been carried forward by Jean Dominique Cassini, his son Jacques (1677–1756) and their collaborators (La Hire, Maraldi) throughout France with works of geodesic triangulation from the north (Dunkerque) to the south down to the Pyrenées (Collioure).

The measurements of the Cassinis and their collaborators in the French territory yielded a result which disproved what Newton's theory maintained: the length of the arc of meridian decreased going northward, supporting the thesis of an elongated spheroid. This fact gave rise to heated debate: on one side, the undisputed authority of the great scientist (Newton); on the other, the experimental results of very accurate measurements performed by authoritative academicians. The circumstance that the Cassinis were Cartesian and thus not followers of Newton's theory of gravitation increased the ardour of the debate. However, their measurements (carried out in 1718 by Jacques Cassini succeeded his father), which suggested an elongated shape for the terrestrial spheroid, were not successful in undermining trust in Newton's calculations.

In order to remove all doubt and resolve the dispute, in 1733 the Académie Royale des Sciences established (by order of King Louis XV) to send two missions, one near the equator and the other near the north pole, to measure the length of the degree of the meridian at latitudes as different as possible in order to obtain more trustworthy results. The first mission left for Peru in 1735, and the second left for Lapland in 1736.

The Académie expended a notable effort (of both men and means), and it received much publicity at that time. In fact it is astonishing that the subject excited such a great interest on the part of the press that even caricatures were published on it (see, for instance, Fig. 5.2).[24]

We must ourselves to giving a bare minimum of details about the two expeditions.

The expedition to Peru was headed by astronomer Louis Godin (1704–1774), mathematician and geographer Charles-Marie de la Condamine (1701–1774), and mathematician and geophysicist Pierre Bouguer (1698–1758), and reached Quito on 29 May 1736. The progress of the measurements was adventurous due to both the intricate orographic features of the country and the disagreements among the

---

[24]The figure is taken from S. Chandrasekhar: *Ellipsoidal Figures of Equilibrium* (1969); rpt. New York: Dover, 1987), p. 3.

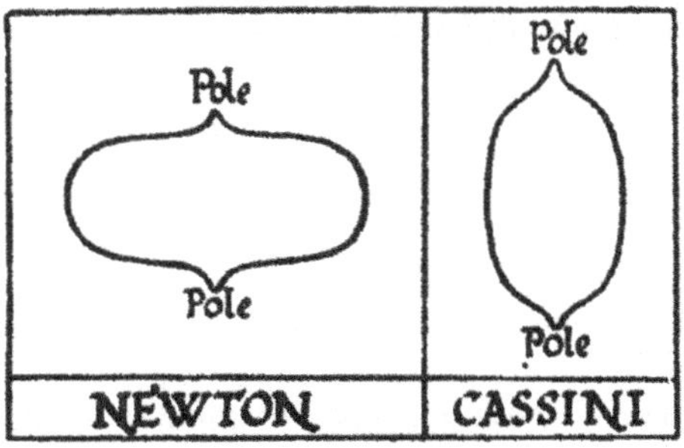

Fig. 5.2 An old-time caricature of the controversy between the opposing schools of Newton and Cassini with respect to the figure of the earth

three scientists who even returned to Paris separately.[25] However, notwithstanding their disagreements, the results of the expedition confirmed Newton's theory: the Earth is a spheroid flattened at the poles.

The expedition to Lapland was headed by Pierre Louis Moreau de Maupertuis (1698–1759), a mathematician and astronomer and follower of Newton. Accompanying Maupertuis were mathematician Alexis-Claude Clairaut (1713–1765), who had already studied the subject and, as we shall see, would later write a fundamental work on the figure of the Earth, mathematician Charles-Etienne-Louis Camus (1699–1768) and astronomer Pierre-Charles Le Monnier (1715–1759), all members of the Académie. The French academicians were joined in Dunkerque by the Swedish astronomer and physicist Anders Celsius. One may be surprised at the ages of the French scientists participating in the mission: the head, Maupertuis, was 38 years old, but Monnier was only 21 and Clairaut only 23 (Clairaut had been promoted to the Académie when he was only 18!).

Among those who followed the progress of the mission with interest was Voltaire, who had promoted Newton's ideas in France. But he was not always on good terms with Maupertuis. In fact, in a first moment, commenting on the results of the expedition, he enthusiastically said that Maupertuis had "*aplati les pôles et les Cassini*" ("flattened the Earth and the Cassinis too"), instead later on wrote the sarcastic verses "*Vous allâtes vérifier en des lieux pleins d'ennuis/ce que Newton trouva sans sortir de chez lui*" (You went to check in places full of trouble/what Newton found without leaving his house).

The Lapland expedition lasted a much shorter time than that in Peru. Maupertuis reported the results at the Académie in the public assembly on 13 November 1737 and then in 1738 published the talk in the book, in which he described all the triangulations.[26]

[25]A novel has been written on the vicissitudes of this mission; see R. Whitaker: *The Mapmaker's Wife: A True Tale of Love, Murder, and Survival in the Amazon* (Basic Books, 2004).

[26]P. L. Maupertuis, *La Figure de la Terre déterminé par les Observations faites par ordre du Roi au cercle polaire – par M. De Maupertuis* (Paris: Imprimerie Royale, 1738).

In the same year 1738, Celsius also published a work concerning the shape of the Earth.[27] This very short work (only 20 pages), rather than being an account of the measurements performed in Lapland, is devoted to a critical examination of the Cassinian observations. The conclusion of this examination is:

> So I hope to have shown, more than enough to a just and impartial reader, that the Cassinian observations, both celestial and terrestrial, mainly performed in the southern France, are too uncertain to be able to infer the Earth's shape from them in any way.[28]

We must add that in the expedition there had also been another, particular member: a priest. He was Abbé Réginald Outhier (1694–1774), corresponding member of the Académie. He kept a diary of the activity of the group and published it in 1744 with the title *Journal d'un voyage au nord.*[29]

One can say that, by the late 1740s, Newton's theory on the Earth's shape was completely accepted in the scientific community, even by the third generation of the "Cassini dynasty", César-Francois Cassini de Thury (1714–1784), known as Cassini IV. We have also dwelt longer on the Lapland expedition because one of its components, Clairaut, will write the first fundamental mathematical work in the modern sense on the Earth.

In conclusion we can recall that the entire subject was recounted in a long article by d'Alembert, in the sixth volume of the *Encyclopédie* (first edition, 1756), with the title *Figure de la Terre*, which began with these words:

> Cette importante question a fait tant de bruit dans ces derniers temps, les Savans s'en sont tellement occupés, sur-tout en France, que nous avons crû devoir en faire l'objet d'un article particulier, sans renvoyer au mot TERRE, qui nous fournira d'ailleurs assez de matiere sur d'autres objets.[30]

5.4 Clairaut's "*Figure de La Terre*"

With the great Clairaut's work, about which we now shall speak, one can say that we are in presence of a symbolic case which confirms the Wigner's assertion of "the unreasonable effectiveness of mathematics in the natural science". This is a

[27]A. Celsius, *De Observationibus Pro Figura Telluris determinanda in Gallia habitis Disquisitio*. Auctore Andrea Celsio. Uppsala: Typis Höjeranis, 1738.

[28]A. Celsius, *De Observationibus*, op. cit. p. 20: *Spero itaque jam aequo & candido Lectori satis superque ostendisse, observationes Cassionianas tam coelestes quam terrestres, in Gallia praecipue meridionali habitas, adeo incertas esse, ut inde figura telluris nullo modo deduci queat* (our Eng. trans.).

[29]*Journal d'un voyage au nord en 1736 & 1737* par M. Outhier, Prêtre du Diocèse de Besançon, correspondant de l'Académie Royale des Sciences (Paris: 1744). The book can be found reprinted in: André Balland: *La Terre Mandarine. Journal d'un voyage au Nord par déterminer la figure de la Terre, par M. l'abbé Réginald Outhier* (Seuil, 1994).

[30]Jean le Rond d'Alembert, *Encyclopédie, ou, Dictionnaire raisonné des sciences, des arts et des métiers*, tome VI (Paris: Denis Diderot, André le Breton, Michel-Antoine David, Laurent Durand, Claude Briasson, 1756), p. 749.

work in which the experimental results (the measurements of the arc of meridian at different latitudes—from the equator to the north pole), that is, "nature", are interpreted by a mathematical theory and the mathematical theory, in its turn, is confirmed by experience. Furthermore, the elaboration of the theory leads to the creation and development of new mathematical tools (in this case, partial differential calculus). Thus the physical problem that must be solved led to the creation, to that end, of "new mathematics", which would then in turn generate further developments; afterwards there would be mathematics generated by mathematics and no longer tied to a particular physical problem.

Clairaut's *Théorie de la Figure de la Terre* was published in 1743[31] and had been preceded by some memoirs published starting in 1735. The work is divided into two parts. The first part bears the title: "General Principles for finding the equilibrium of fluids and for determining the shape of the Earth and the other planets when the law of gravity is given". Chapter 4 of this first part assumes supreme importance, since it is here that Clairaut enunciates the mathematical conditions required for the equilibrium of fluids.

He makes use of the artifice introduced by Newton (that of channels), but in a more general way. He sets the problem thus (p. 33): Given the law with which gravity acts on the points of a fluid mass rotating around an axis, find if this mass can maintain a constant shape.

He refers to Fig. 5.3 (a quarter of a meridian), where ON is a channel in the plane of the meridian CEP and maintains that the necessary and sufficient condition for a fluid mass to be able to assume a constant shape is that any closed channel in the plane of a meridian must be in equilibrium independently of the centrifugal force. That is, it is necessary and sufficient that, when one calculates the total effort of gravity in any channel ON, one obtains the same quantity that one would obtain considering any other channel joining O with N.

Let us consider S and s infinitely close and set CH = x, HS = y, Sr = dx, sr = dy. In addition let us call P and Q the components of the force of gravity at any point S, the one normal to CP and the other parallel. Pdx + Qdy will be the total effort on the circular cylinder Ss due to gravity.[32]

Since the line integral of this differential along the channel between any two points must not depend on the line but only on the two extremes, the differential will have to be an exact differential, i.e., it will have to be $\frac{\partial P}{\partial x} = \frac{\partial Q}{\partial y}$.

---

[31] A.-C. Clairaut, *Théorie de la Figure de la Terre, tirée des Principes de l'Hydrostatique*... (Paris: David Fils, 1743). The book was reprinted in 1808, edited by Denis Poisson. Only two translations from the French are available: (German) *Theorie der Erdgestalt nach Gesetzen der Hydrostatik* herausgegeben von Ph. E. B. Jourdain und A. v. Oettingen (Leipzig, 1913); (Italian) *Teoria della Forma della Terra dedotta dai principi dell'Idrostatica*, traduzione e note di M. Lombardini seguite da una nota di F. Enriques (Bologna: Nicola Zanichelli, 1928).

[32] The reader can find a detailed discussion on this point (and also on the subsequent ones) in Greenberg's *The problem of the Earth Shape* ..., op. cit., in which about 200 pages are devoted to the Clairaut's work. We must limit ourselves to reprising the results.

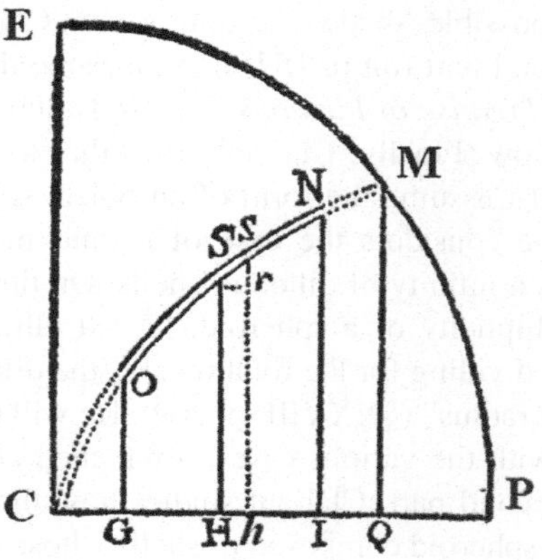

Fig. 5.3 Clairaut's figure for determining the shape of the Earth. From A.-C. Clairaut, *Théorie de la Figure de la Terre*, p. 34

At this point Clairaut maintains: Each time this equation is satisfied, it is certain that there is equilibrium in the fluid.

Considering the closed channel CQM, one see that the total effort of the centrifugal force per unit of mass on the straight channel CQ is equal to zero because CQ is parallel to the rotation axis. Therefore

$$\int_C^M \frac{fy}{r}dy = \int_Q^M \frac{fy}{r}dy,$$

where *f* designates the centrifugal force per unit of mass at a distance *r* from the axis of symmetry of the homogeneous figure of revolution in equilibrium. Then the centrifugal effort of the column CM will be $\frac{fy^2}{2r}$, and consequently the general equation of the meridian will be

$$\int_C^M Pdy + Qdx - \frac{fy^2}{2r} = A \quad \text{(a constant independent of M).}$$

Whereas in the first part of the work Clairaut also takes into consideration the theories of the "Cartesians" (i.e., of Huygens), in the second part he makes use exclusively of Newton's theory for the gravitational attraction. In fact, it bears the title "Determination of the shape of the Earth and of the other planets, by assuming that all their parts attract mutually with the inverse square law of distances".

He begins by considering the case of homogeneous planets, proving the proposition that Newton has only asserted,[33] i.e., that for a planet the shape of an

[33]A.-C. Clairaut, *Theorie de la Figure de la Terre*, p. 153: *Cependant M. Newton, bien loin d'avoir démontré que la terre était ou approchait d'être un sphéroïde elliptique, n'a pas seulement affirmé qu'elle l'etait.*

oblate spheroid is a possible shape of equilibrium. Clairaut points out in the introduction to the second part (on p. 157) that the same theory was also given by Maclaurin in his work *Treatise of Fluxions*, but, as Todhunter notices,[34] Clairaut's version is easier to follow. Finally, Clairaut holds that a rotating mass of fluid in relative equilibrium must assume the form of an oblate spheroid (§IX, p. 171).

Next, in Chap. 2, he considers the case of a fluid mass which coats a solid spheroid composed of an infinity of different shells. On this occasion he also gives the definition of the ellipticity of a spheroid: "I call ellipticity of a spheroid the fraction one obtains by dividing for the rotation axis the difference between the axis itself and the equatorial radius" (§XXVIII, p. 209). He will use the magnitude of the ellipticity for dealing with the various cases of flattened ellipsoids.

In Chap. 3 of the second part, Clairaut studies how the weight varies from the equator to the pole in a spheroid composed of shells whose densities and ellipticities vary arbitrarily from the centre to the surface. He deals with the relative equilibrium of rotating fluids supposed to be ellipsoidal and rotating around a common axis, and nearly spherical. The fundamental result of this chapter is that which is known as Clairaut's theorem. If, in general, we define the ellipticity of a rotating ellipsoid with $=\frac{a-b}{b}$, where a is the equatorial radius and b the rotation semi-axis, and $v=\frac{P-\Pi}{\Pi}$ the "Clairaut's fraction", where P is the weight at the pole and Π the weight at the equator, Clairaut's theorem holds that $\frac{P-\Pi}{\Pi}=2\varepsilon-\delta$, where ε is the ellipticity in the particular case of homogeneous ellipsoid. For a homogeneous spheroid it also holds $\varepsilon=\frac{5}{4}\phi$, where ϕ is the ratio between the centrifugal force and the weight at the equator. The ratio $\frac{a-b}{a}=s$ is the flattening, which in Clairaut's approximations is $\sim\delta$.

In Chap. 4 the Earth's shape is studied in the case of an initial fluid state resulting from an infinity of fluids of different densities. The result is that, if one supposes the planet (Earth) to have been fluid and non-homogenous at the beginning, then:

(1) the weight must decrease on the surface from the pole to the equator by an amount more significant than if it had been homogenous;
(2) this decrease is proportional to the square of the cosine of the latitude.

Finally the amount by which the weight at the pole exceeds the weight at the equator is given by $2\varepsilon-\delta$. Clairaut devoted the closing chapter to a comparison of his theory with the observations. The measurements performed in Lapland were not yet sufficiently reliable for a rigorous determination of the terrestrial flattening and he was waiting for the results from Peru for calculating more exactly its value, which afterwards came out in accord with the theory.

In the natural course of things, after Clairaut's work, many contributions appeared regarding both fluid dynamics and the Earth's shape. The most significant names, in the second half of the eighteenth century, are those of D'Alembert, Legendre and Laplace. D'Alembert, in addition to the already quoted article in the

[34] I. Todhunter, *A History of the Mathematical Theories of Attraction*, op. cit. p. 203.

Encyclopédie, published many memoirs on the subject (the last of them in 1780). Laplace devoted to the subject many contributions, also in his treatise *Mécanique Céleste* up to the fifth volume published in 1825. We shall not deal with these contributions, since our purpose is to remark the fundamental junctions, both in the history of thought and in the experimental results, concerning the shape (and the size) of the Earth, and these contributions obviously improved the theory but not changed it. As Todhunter wrote (in 1873):

> In the Figure of the Earth no other person has accomplished so much as Clairaut; and the subject remains at present substan-tially as he left it, though the form is different. The splendid analysis which Laplace supplied adorned, but did not really alter, the theory which started from the creative hands of Clairaut.[35]

5.5 The Metre

Until now we have tried to follow, at least in their outline, besides the theoretical developments, also the most important experimental researches concerning the shape and the size of the Earth. As we noted, a problem that in time had caused complications was that of being able to compare the measures of length performed in different nations and with different units of length (every nation had its own units of standard of measurement, just as it had its own money). As a consequence, among the geodesists of the eighteenth century, the necessity of having everywhere a sample of precise and reliable length that was easily reproducible in principle soon became pressing. As we have seen, a cultural environment that felt this need especially keenly was the French one, for both scientific and institutional and administrative purposes. (The English institutions were interested in the matter, but at the end they decided not to collaborate with the French.)

In France, instead the interest was so great that the problem was taken into consideration even at a time when the French Revolution had been underway for a year. In fact, following a proposal made by Charles-Maurice Talleyrand (1754–1838), in 1790 the Assemblée Constituante appointed a committee consisting of Jean-Charles Borda (1733–1799), Joseph Louis Lagrange (1736–1813), Pierre Simon Laplace (1749–1827), Gaspard Monge (1746–1818) and Jean-Marie Antoine Nicolas Caritat de Condorcet (1743–1794) (in hindsight, it would have been very hard to choose a more prestigious committee!), in order to elaborate a proposal for the choice and the definition of a new unit of length, which was to be both natural and universal.

The committee presented a report on 19 March 1791, suggesting two options for unifying the measures of length: the length of a seconds pendulum (at 45° of latitude and at the sea level and at the temperature of melting ice), or the ten millionth part of the quarter of the terrestrial meridian from Dunkerque to

[35] I. Todhunter: *A History of the Mathematical Theories of Attraction …*, op. cit., vol. I, p. 229.

Barcelona.[36] It can be observed that these two units were not actually very different with regard to their length: the pendulum's length was close to that of the second choice.

On 26 March 1791 the Assemblée Constituante adopted this report and King Louis XVI (still reigning) instructed the Académie to appoint the commissioners for implementing it.

Before continuing our discussion, let us recall what conditions the new unit of measurement had to satisfy, according to the French academicians:

- The new unit was not to be national, so that it could be accepted universally.
- It was not to vary in space and in time.
- It was to allow the construction of easily reproduced samples.
- It was not to have a value too different from the values of units already existing.
- It was to have a really new name, different from the already existing ones.
- To avoid nationalisms, it was to depend exclusively on natural phenomena.

These conditions were well evident and shared by all the people involved. These conditions were also supported by the ideas defended by the Revolution.

Since the length of the seconds pendulum was tied to place and also to the definition of the second (consequently of the day), the length of the quarter of the meridian was preferred. Furthermore, for that length it was possible to construct all necessary samples valid throughout the world. The new unit was called "metre".

It was decided to repeat the measurements of the meridian, since the existing measures were considered rather inaccurate. The commissioners selected by the Académie had been astronomer Cassini IV, mathematician Adrien Marie Legendre (1752–1833) and astronomer Pierre Méchain (1744–1804). It did not take Cassini IV and Legendre long to back out; they were replaced by the young astronomer Jean-Baptiste Delambre (1749–1822).

It is interesting to read Delambre's opinion on this fact in his work, published posthumously, *Grandeur et figure de la Terre*:

> Les commissaires nommés dans l'origine, pour partager avec Méchain les travaux de la Méridienne, étaient Cassini et Le Gendre. A l'époque où les opérations allaient commencer ils s'excusèrent l'un et l'autre; je venais d'entrer à l'Académie et ils consentaient a s'en reposer sur moi. Je n'avais rien a objecter a Le Gendre: je crojais étre sur que jamais il n'avaìt eu l'idée de se charger de cette opération. Cassini n'avaìt aucune excuse réelle, sinon les opinions politiques, qui ne lui permettaient aucun rap-port avec un gouvenement qu'il ne voulait pas reconnaitre; mais ce motif, quelle que soit la force qu'on lui suppose, n'était pas de nature à ètre mis en avant. Je trouvais déjà que c'était une impru-dence assez grande, dans les circonstances où nous nous trouvions alors, que de refuser une mission qui, en lui faisant peut-étre courir quelques risques, le preservait de périls plus imminents et auxquels il n'a pas longtemps échappé. Je ne craignais ni ne dési-rais cette mission qu'il

[36]Actually the suggested options had been three, since the measure of the equator was also suggested, but this was immediately discarded.

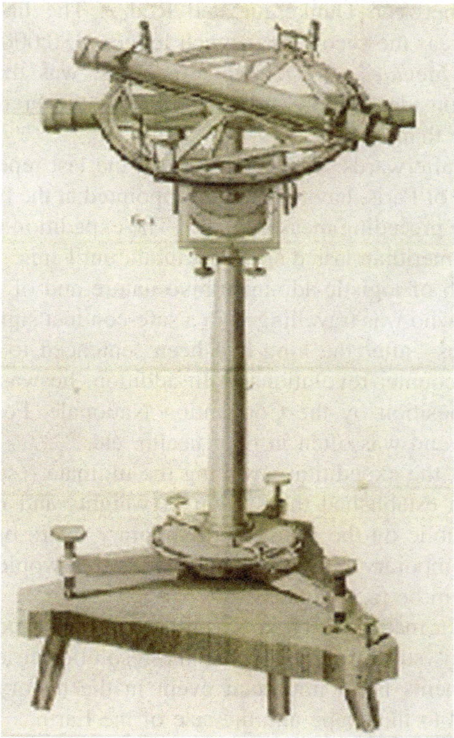

Fig. 5.4 The Borda repeating circle

refusait; je fis tous mes efforts pour lui faire changer sa résolution: il fut inébranlable et je fus aussitòt designé pour remplacer Le Gendre et Cassini.[37]

In this case we think it superfluous to supply an English translation.

More than a year passed after that resolution, while the preparation of the necessary instruments built on purpose for those measurements went forward. One of these instruments (four copies of it were needed), invented by the assistant of Borda (Lenoir) and perfected by Borda himself, was the repeating circle equipped with two telescopes set up on a common axis at the centre (Fig. 5.4).

Other important instruments were those equipped with parabolic mirrors for reverberating the light to be used when the beacons were not sufficiently efficient. Finally on 25 June 1792 Méchain set off, with the first two repeating circles.

According to the established agreement, Méchain was to have measured the part of the meridian between Rodez (in the south of France) and Barcelona, and

[37]J. B. J. Delambre: *Grandeur e Figure de la Terre*, ed. G. Bigourdan (Gauthier-Villars, 1912), pp. 204–205. In this work, published 90 years after his death, Delambre recounts incidents and makes explicit the names which he had omitted in the volumes of the *Base du système métrique décimal* (see infra), perhaps for the sake of peace.

Delambre the part between Dunkerque and Rodez. The first was estimated at 170,000 toises, whereas the second was much longer (380,000 toises). The reason of this "*repartition inégale*", as Delambre recounts, was that the Spanish part (wholly new) was considered more arduous, whereas the French one had already been measured more than once.

Delambre set off afterwards. While waiting for the last repeating circle, he had explored the vicinity of Paris, having been disappointed at the bad conditions of the structures used in the preceding measurements. The expedition for the measurement of the quarter of the meridian lasted six years in all, until June 1798, and was beset with difficulties, both of logistic-administrative nature and of personal origin. For instance, Delambre, who was travelling with a safe-conduct signed by the king, was stopped several times—after the king had been sentenced to the guillotine—and accused of being a counter-revolutionary. In addition, he was, for a long period, removed from his position by the Convention Nationale. For his part, Méchain suffered an accident and was often in poor health, etc.[38]

Before the end of the expedition, awaiting the ultimate results, the Convention passed an act which established the system of weights and measures within the territory of the Republic on the basis of a temporary metre based on the toise of Peru, as well as a temporary kilogram. Only afterwards would the true, definitive decimal metric system be promulgated.

Obviously this is a matter which is beyond the scope of our investigation. We have talked about this subject only because the resolution to establish a universal system of measurements is an important event in the history of science, and is intimately connected to the shape and the size of the Earth.

As always happens in all the experiences of measurement, time led successively to further improvements so that, once a metal sample for the metre based on the measures of Delambre and Méchain had been constructed, the measurements effected in the nineteenth century induced several scholars to compile tables with the "errors of the metre".

To end our narration, we tell that it rested with Delambre to write the mammoth work in three volumes (published in 1806, 1807, 1810)[39] containing the account of the operations carried out and of the mathematical speculations put into effect.

Méchain died in 1804, but he managed to send Delambre the account of his measurements from Rodez to Barcelona (pp, 289–510 of the first volume). Those of Delambre (Dunkerque-Rodez) occupy pp. 1–288. In the remaining two volumes, Delambre reports the rest of observations and calculations.

---

[38]The several adventures in the expedition are narrated in the novels by Denis Guedj: *La Méridienne*, first French edition, 1991 (English edition with the title *The Measure of the World*, University of Chicago Press, 2001) and by K. Adler: *The Measure of All Things: The Seven-Year Odyssey and Hidden Error that Transformed the World* (The Free Press, 2002).

[39]J. B. J. Delambre: *Base du Système Métric Décimal ou Mesure de L'Arc du Méridien compri entre les Parallèles de Dunkerque et Barcelone, exécutée en 1792 et Annés suivantes par mm. Méchain et Delambre*, 3 vols. (Paris: Baudoin, 1806–1807–1810).

If we were to restrict ourselves to the elements recounted up to this point, it would seem that we are only dealing with an umpteenth measurement of the terrestrial meridian, destined never to be the ultimate one, and with the decision of the institutions to create a new system of units of measure of length (afterwards extended to weight). However, what actually lay behind this was all the thought of the Enlightenment, the uninterrupted activity of the Académie, and most of all the work of the greatest mathematicians of that time, destined to stand proud in the history of mathematics.

That period truly witnessed an exceptional scientific development, since the problems suggested in practical terms by the needs of society (in that situation, particularly geodesy) were in the same moment studied by both mathematicians and physicists, and with the collaboration of the best of technology (we have seen the example of the repeating circle). The problems suggested by geodesy, which we have already come across in the Clairaut's work (attraction of the ellipsoids, equilibrium of rotating bodies), extended to potential theory, to the geometry of surfaces (the future differential geometry) and then, in particular, to spherical trigonometry and its generalization in spheroidal trigonometry.

To the already mentioned names of Legendre, Laplace, Monge, we can add Jean-Baptiste Meusnier (1754–1793) and the Swiss Leonard Euler (1707–1783), for the theory of surfaces, aware that they only are the highest peaks of a chain of mountains.

Suggested Readings

Chauvin, L. *Histoire du Metre. D'Après les Travaux et Rapports de Delambre, Méchain, Van Swinden etc*. Rpt. Forgotten Books, 2017.

Diderot, D. (1754). *Pensées sur l'interprétation de la nature*. (Modern edition: ed. Flammarion, 2009).

Hankins, T. L. (1970). *Jean D'Alembert: Science and Enlightment*. Oxford: Clarendon Press.

Kellogg, O. D. (1929). *Foundation of potential theory*. Rpt. New York: Dover 1954.

Laplace, P. S. (1824). *Exposition du système du monde*, 5th ed. Paris: Bachelier.

Rossi, P. (2001). *The Paolo Rossi: The birth of modern science*. Blackwell.

Voltaire. (1738). *The elements of Sir Isaac Newton's Philosophy translated from the French*. Revised and corrected by John Hanna. London, Stephen Austen, (rpt. Gale ECCO, 2010).

Chapter 6
From the French Revolution to the Artificial Satellites

6.1 Towards the Geoid

Napoleon, upon receiving from the hands of Delambre the volume of the *Base du Système métrique décimal*, congratulated the author saying "*Les conquêtes passent mais ces opérations restent!*". This laudatory sentence obviously did not imply an encouragement to consider oneselves definitively satisfied with the obtained results. In fact, starting in the first years of the new century, there was a movement to repeat, with renewed accuracy and with improved equipments, the measurements executed in the past.

In Sweden, in 1801, astronomer Daniel Melanderhjelm (1726–1810) planned to repeat Maupertuis' measurements in Lapland. In view of his old age, he assigned the task to Jöns Svamberg (1771–1851) who, together with a few collaborators,[1] repeated the survey made by Maupertuis, extending it southwards and northwards to a total length of 1° 37′. We know (from Delambre's *Grandeur et Figure de la Terre*, p. 318) that Melanderhjelm also had asked Delambre, with whom he was in correspondence, to send him a repeating circle made by Étienne Lenoir (1744–1832) for better executing the triangulations. The final judgement of Delambre on the measurement was only partly positive: "*On pourrait dire en général qu'on y reconnait le géomètre estimable plus que l'astronome bien exercé*".[2]

The resolution to repeat, or in any case to enlarge the measurements made in the past also regarded the "*Méridienne de Paris*", i.e., the measurement of the arc performed by Delambre and Méchain. We know that Méchain would have liked to continue the measurements as far as the Balearic Islands but he was prevented from doing this by various misfortunes, and ultimately by his death.

[1]See the account of the measurements: *Exposition des opérations faites en Lapponie pour la determination d'un Arc de Meridien* (1801, 1802 et 1803) par MM. Ofverbom, Svanberg, Holmquist et Palander, Redigée par Jöns Svamberg… (Academie royale des Sciences de Stockholm, 1805).

[2]J. B. J. Delambre: *Grandeur et Figure de la Terre*, op. cit., p. 330.

D. Boccaletti, *The Shape and Size of the Earth*,
https://doi.org/10.1007/978-3-319-90593-8_6

Thus it happened that Laplace managed to obtain a mandate to allow the young François Jean Dominique Arago (1786–1853), together with Jean Baptiste Biot (1774–1862), to terminate the measurements of the geographic meridian interrupted by Méchain's death. Arago and Biot left Paris in 1806 and started the operations along the mountains of Spain. Biot returned to Paris after having determined the latitude of Formentera in the Balearic Islands. Arago instead continued the measurements but had a series of mishaps (he was mistaken for an informer by the Spaniards and imprisoned; he managed to escape but was then imprisoned again by pirates; he was definitively released in 1809).[3] Luckily, he was able to hold on to the documentation of his measurements and deposited it at the Bureau de longitudes in Paris. For his merits he was elected member of the Académie des Sciences (at the age of only 23!). Here again the arc had been extended (to 2° 42′). As we shall see presently, the idea of performing new triangulations afterwards spread throughout Europe and, later, even to the United States.

We have dwelt on the French measurements to emphasize the unflagging interest of the Académie and the Observatoire de Paris in perfecting their results (after all, their measurements had led to the definition of the metre and of the decimal metric system!).

We take the opportunity to recall that the decimal metric system was never accepted by the historically English-speaking nations. Their unit length, the yard, also had a long history of definitions. Among them, one of the last (actually only an attempt) was the proposal (examined but not approved) of a committee of the English parliament in 1814 to base the definition of the yard upon the length of a seconds pendulum. *Nihil sub sole novum*! Finally in 1959, by an agreement between the United Kingdom, the United States and other nations, the yard was legally defined as being equal to 0.9144 m.

But the definition of the metre, even though remaining the length of the original sample, in the last decades, has also been subject to transformations in order to disengage it from a comparison with the original metal bar. The present definition is 1 m = the length of the path travelled by light in vacuum during a time interval of 1/299792458 of a second. In its turn, the second is now defined as the duration of 9192631770 periods of the radiation corresponding to the transition between the two hyperfine levels of the ground state of the cesium 133 atom.

Let us return now to our subject. It is impossible for us to mention all the operations that were executed in all Europe in the nineteenth century. The thing which must be emphasized is that over the course of time most governments realized that the surveying of their own territory was an impelling necessity for the national progress. Thus new institutions were born, new instruments and new procedures were studied. For instance the number of bases were increased (often

[3]Arago narrated all this in detail in his autobiography; see: *Histoire de ma Jeunesse* par François Arago, precedée d'une Preface par Alex. De Humboldt (Bruxelles & Leipzig: Kiessling, Schnée, 1854).

three, whereas before only one was used), and when the wireless telegraph came into use communications between the stations improved significantly.

Among the first countries to undertake campaigns of measurements was England, which in 1783 began, under the guidance of Major-General William Roy, the triangulation which was continued under the ensuing direction of the Ordnance Survey. The principal meridian chain extended from the Isle of Wight to the Shetland Islands over a length of 10° 13′. Since 1860 geodetic work of extreme importance was also undertaken, thanks to the co-operation of the governments of Prussia, Russia, Belgium, France, and England, in order to connect the triangulation of France, Belgium, Russia, and Prussia, which were sufficiently advanced for that purpose in 1860, with the principal triangulation of Great Britain (including Ireland), which had been finished in 1851. It is interesting to quote the considerations of a protagonist of these measurements (the general Sir Henry James):

> Before the connexion of the triangulation of the several countries into one great network of triangles extending across the entire breadth of Europe, and before the discovery of the Electric Telegraph, and its extension from Valentia (Ireland) to the Ural mountains, it was not possible to execute so vast an undertaking as that which is now in progress. It is, in fact, a work which could not possibly have been executed at any earlier period in the history of the world. The exact determination of the Figure and Dimensions of the Earth has been one great aim of astronomers for upwards of two thousand years; and it is fortunate that we live in a time when men are so enlightened as to combine their labours to effect an object which is desired by all, and at the first moment when it was possible to execute it.[4]

At the beginning of the nineteenth century measurements of both latitude and of longitude were also undertaken in India ("The Great Trigonometric Survey of India", 1800–1802) between Madras and Mangalore. Afterwards Lieutenant Colonel William Lambton (1753–1823) continued the surveys northward along a meridian chain of more than 20° for about twenty years. He died during the survey in the central India and was succeeded by his assistant George Everest (1790–1866), for whom Mount Everest was named in the English language by the Royal Geographic Society). In those years (1816–1865) the Russo-German Friedrich George Wilhelm von Struve (1793–1864) also undertook a chain of triangulations stretching from Hammerfest in Norway to the Black Sea, passing through ten nations and along 2820 km. The so-called Struve Geodetic Arc has been in the UNESCO list of World Heritage Sites in Europe since 2005. The measurements ended in 1855 and provided an arc of a length of 25° 20′; the flattening of the Earth was estimated to be 1:294.26.

As we have hinted above, large scale networks (chains and filling nets) were also developed in the USA starting in 1830. These researches culminated in the twentieth century in the work of John Fillmore Hayford (1868–1925), who devoted himself to the construction of the reference ellipsoid and in 1909 determined two values: a = 6,378,388 m for the semi-major axis, and 1:297.0 for the flattening of

[4]*Comparisons of the Standards of Length of England, France, Belgium, Prussia, Russia, India, Australia, made at the Ordnance Survey Office, Southampton, by Captain A. E. Clarke, R. E., under the direction of Colonel Sir Henry James, R. E., F. R. S.* Published by order of the Secretary of State for War, 1866.

the Earth (these values were adopted in 1924 as the International Ellipsoid of Reference, also known as the Hayford ellipsoid).

Let us turn now to deal with the researches carried out in Europe, particularly in pre-unification Germany. The first very important survey from the scientific point of view was Hannover arc measurement (1821–1823) by Carl Friedrich Gauss (1777–1855) followed by another local, but outstanding, geodetic contribution due to Friedrich Wilhelm Bessel (1784–1846), professor at University of Köninsberg in Eastern Prussia. Like Gauss, Bessel took the view that their measurements should join the other European efforts to determine of the shape of the Earth.

While listing the various enterprises concerning central Europe, we must recall that the continent was divided into various small states and that Germany was unified only in 1871, becoming the German Empire with Otto von Bismarck as founder and first chancellor. Moreover, in the nineteenth century Europe was the theatre of several wars. The persistence in projecting and executing campaigns of measurements in spite of all the unrest interesting the whole of Europe is evidence that all governments were convinced of the importance and the necessity of these measurements.

We continue now with the measurement projected and executed by German scientists.[5] First, let us recall that since the time of Laplace (1808) it had become evident that a model consisting in a rotational ellipsoid could no longer represent the Earth with the necessary accuracy. Further, both Gauss (1808)[6] and Bessel (1837) noticed that the deviation of the physical plumb line from the normal to the ellipsoid (the so-called "deflection of the vertical") could no longer be ignored, and thus began the search for a new "ideal surface" that more closely matched the physical surface of the Earth everywhere. The ellipsoid, itself no longer the ideal, would instead become a reference frame approximating it.

The great campaign in Europe was initiated by the retired Prussian General Johann Jacob Baeyer (1794–1885) in 1861. Taking up the earlier ideas of Gauss, Bessel, Struve and others, he proposed an arc measurement project for central Europe. In his project the network of triangulations would range north-south from southern Italy to Norway and west-east from France to Poland. The project immediately obtained the support of the Prussian government and, in spite of the fact that at that time Europe was not without conflicts, the participation of the European countries surpassed all expectations. Sixteen governments declared their participation, including seven German states and others from Sweden and Norway to Italy. Obviously this fact had as a consequence the necessity of an efficient organization. Baeyer handled the situation by relying on his previous military career, during which he had collaborated on various campaigns of measurements.

---

[5]For a complete information on this subject we refer the reader to W. Torge: "From a Regional Project to an International Organization: "The Baeyer-Helmert-Era" of the International Association of Geodesy 1862–1916", in Chris Rizos, Pascal Willis, eds., *IAG 150 Years*. International Association of Geodesy Symposia, vol 143. (Cham: Springer, 2015), pp. 13–18.

[6]C. F. Gauss (1828): *Bestimmung des Breitenunterschiedes zwischen der Sternwarten von Göttingen und Altona*, in Carl Friedrich Gauss, *Werke*, vol. IX (Leipzig: Teubner, 1903) (in German).

Among the objectives of the project was the determination of the deflection of the vertical and so with the inclusion of "gravity measurements" in the program of the European Arc Measurement, this physical tool again entered the realm of geodesy. In the years in which the project was carried on, many technical improvements were introduced in which mathematical tools were also essential. For instance, the multiplication of the number of bases, and the necessity of ensuring that the values of the angles of the triangles obtained from a direct measurement also satisfied the condition of the correct sum of the internal angles of each triangle caused a superabundance of data from which the most reliable values were to be deduced. To that end the so-called least squares method, due to Gauss, was systematically applied, making it possible to achieve the most reliable result while minimizing the sum of squared residuals.[7] We also recall that, even if at the outset the project was hampered by the usual confusion of units of length, finally in 1867 the metre was recommended.

In the years in which Baeyer was working on his project, it also came across quite clearly what form was assumed by the ideal surface representing (locally) the physical surface of the Earth more faithfully than a rotational ellipsoid. This surface, named "geoid" in 1873 by Johann Benedict Listing (1808–1882), a disciple of Gauss, can be defined essentially as the equipotential surface of the Earth's gravity field coinciding with the mean sea level of the oceans. That surface is imagined as being obtained by extending (ideally) the surface, in a state of equilibrium, of the oceans under the continents. As Wolfgang Torge says "The death of Baeyer finishes the first epoch of organized international collaboration in scientific geodesy".[8]

In the meantime, in unified Germany a scientist came to the fore who wished to see modern geodesy founded as a proper science, autonomous and no longer shared with astronomers, mathematicians and military academies. This is Friedrich Robert Helmert (1843–1917), who in 1880 published the first part of his treatise (the second part came out in 1884),[9] in which he defines geodesy: "Geodesy is the science of measuring and mapping the Earth's surface".[10] We can also add that, after Baeyer's death, the collaboration of the European Arc Measurement was

---

[7]Although one can find a first version of this method in a work of Legendre (1805) on the determination of the orbits of comets, the method is usually credited to Gauss. He expounded it in his work *Theoria Motus Corporum Coelestium in sectionibus conicis solem ambientium*, 1809, (a many times reprinted English translation "Theory of the motion of the Heavenly Bodies moving about the sun in conic sections" can be found in a Dover edition, 2004). For a treatment of the subject in the spirit of Gauss work we refer the reader to a book published in 1892 (but still today reprinted): *Theory of Errors and Method of Least Squares* by William Woolsey Johnson.

[8]Torge: "From a Regional Project", op. cit., p. 12.

[9]F. B. Helmert: *Die Mathematischen und Physikalischen Theorien der Höheren Geodäsie*, 2 vols. (Mathematical and Physical Theories of Higher Geodesy) (Leipzig: Teubner, 1880/1884). An English translation is: F. R. Helmert, (1964). Mathematical and Physical Theories of Higher Geodesy, Part 1, Preface and the Mathematical Theories (St. Louis: Aeronautical Chart and Information Center, 1964), available at http://doi.org/10.5281/zenodo.32050.

[10]F. B. Helmert: *Die Mathematischen und Physikalischen Theorien*, op. cit., vol. 1, p. 3: *Die Geodäsie ist die Wissenschaft von der Ausmessung und Abbildung der Erdoberfläche* (Eng. trans. Aeronautical Chart and Information Centre).

transformed into an international organization in which North and South America and Japan also were involved. Over the course of the nineteenth century, starting from the Méridienne de Paris, measuring had even arrived in Japan! In this organization an outstanding position (Director of "Central Bureau") was held by Helmert himself, who would also become the President of the Prussian Geodetic Institute.[11] Helmert was certainly the most important figure in the field of geodesy at the European level up to World War I.

Helmert also realized how important the determination of the Earth gravity field was for the science of geodesy and for the other earth sciences as well. The gravity measurements were particularly important for establishing, also through the observed deflections of the vertical, the local features of the geoid. To that end, the manufacturing of transportable pendulums was notably improved. Nevertheless, the triangulations for the computation of reference ellipsoids continued (above we have recalled the results of Hayford).

As our aim is not to narrate the history of geodesy but, as we have many times noted, to point out how the problem of determining the shape and size of the Earth has coincided with junctures in the development of the philosophical and scientific thought in the history of humanity, we will go no further with describing the results of measuring campaigns. These obviously continued, with the improvement of the traditional techniques thanks to the progress of technology. Only when the artificial satellites entered the field did the situation changed.

6.2 The Geoid and the End of Our Journey

We consider a fundamental finishing line, both experimental and theoretical, the acceptance of the notion that the figure of the Earth can be approximated, and consequently described, only locally and sometimes only for a certain time, by a well-defined geometrical surface. That surface was first conceived as the sphere (the perfect ideal solid), then ideas about it shifted to the rotating ellipsoid with axes measured with great accuracy. Later even that, at first considered as the "definitive" surface, was eventually demoted to the rank of reference frame.

However, before going on to speak about the geoid and the mathematical problems concerning the geometry of reference frame, let us formulate a more general discourse articulated in various directions. In the nineteenth century, while the measurements about which we have spoken were executed, new branches of mathematics and physics showed an impetuous growth. Some of those branches bore a close connection with problems related to the theories about the figure of the Earth.

We have seen that, for the first time with Clairaut (and Maclaurin), there was a desire to determine the steady shape of a rotating solid—the Earth—held together by the force of gravity in accordance with Newton's law. Clairaut's work was only

[11] See: Johannes Ihde and Andreas Reinhold, "Friedrich Robert Helmert, founder of modern geodesy, on the occasion of the centenary of his death", *History of Geo- and Space Sciences* 8 (2017): 79–95.

a commencement. The subject of a rotating ellipsoid, in its various representations, became a field of research which gave birth to further developments in different branches of mathematics. After the writings of Laplace and Poisson where Maclaurin's ellipsoid was still considered the only admissible solution to the problem of equilibrium of uniformly rotating homogeneous masses, Lagrange in the first volume of the second edition of his *Mécanique Analitique* (published in 1811) gave an (apparently) rigorous solution to the problem. According to Lagrange, in a rotating ellipsoid at equilibrium the two equatorial axes must necessarily be equal.[12]

It was Carl Gustav Jacob Jacobi (1804–1851) who recognized in 1834 the inadequacy of Lagrange's demonstration, proving:

> that ellipsoids with three unequal axes can very well be figures of equilibrium; and that one can assume an ellipse of arbitrary shape for the equatorial section and determine the third axis (which is also the least of the three axes) and the angular velocity of rotation such that the ellipsoid is a figure of equilibrium.[13]

As an example of how broad the problem of the ellipsoid of equilibrium had become, we can cite the papers of Edmond Roche (1820–1883) who, in the years 1849–1851 considered the problem of the equilibrium and the stability of rotating homogeneous masses which are, further, distorted by the constant tidal action of an attendant rigid spherical mass. He would then describe a method of calculating the distance at which an object held together only by gravity would break up due to tidal forces; this distance became known as the Roche limit. Afterwards, in the twentieth century, the model of Roche's ellipsoid would be applied in stellar dynamics.[14]

After Lagrange and Jacobi, other great mathematicians, among them Dedekind, Dirichlet and Riemann, dealt with the subject.[15] Towards the end of the century Henri Poincaré (1854–1912) also devoted himself into the problem. He began with a short communication in the *Comptes Rendus de l'Académie des Sciences* in 1885, where he gave a proof of a proposition regarding annular masses, which had been considered valid without proof in a famous text at that time.[16] But what drew the attention of the readers was the statement he made as conclusion:

> *Je n'ai pu encore approfondir la question de la stabilité de ces masses annulaires. J'ai fait seulement, en passant, une remarque que je crois nouvelle.*

---

[12]In the third edition (1853), the passage in question was commented by a note of the editor J. Bertrand. See: *Mécanique Analitique*, par J. Lagrange, Troisième edition, Revue, Corrigée et Annotée par M. J. Bertrand (Paris: 1853), p. 192, 405–407. The note of Bertrand mentioned the successive development of the theory due to Jacobi.

[13]Quoted in S. Chandrasekhar: "Ellipsoidal Figures of Equilibrium—An Historical Account". *Communications on Pure and Applied Mathematics* 20 (1967): 251–265.

[14]See, for instance: W. C. Saslaw: *Gravitational Physics of Stellar and Galactic Systems* (Cambridge University Press, 1985); S. L. Shapiro, S. A. Teukolsky: *Black Holes, White Dwarfs, and Neutron Stars* (Wiley, 1983).

[15]See for instance, S. Chandrasekhar: *Ellipsoidal Figures of Equilibrium*, op. cit.

[16]It was the second edition of the famous textbook *Treatise on Natural Philosophy* by William Thomson (Lord Kelvin) and P. G. Tait, which was successively entitled *Principles of Mechanics and Dynamics* and had a last revised edition in 1912.

> *Si la vitesse angulaire ω est supérieure à $\sqrt{2\pi}$ (avec les unités adoptées) il n'y a plus pour la masse fluide aucune figure d'équilibre stable possible.*[17]

Thus Poincaré maintains that for any figure of rotation the condition holds: $\omega > \sqrt{2\pi f\rho}$, where ω is the velocity of rotation, ρ the density of the solid and f the constant of gravitation. Immediately after he proved that condition, and it can be found as "*Theorème de Poincaré donnant pour vitesse de rotation une limite au delà de laquelle l'équilibre est impossible*" in the *Traité de Mécanique Céleste* by François Tisserand, where the second volume is wholly devoted to the figure of celestial bodies and their movements of rotation.[18]

Poincaré's effort to study the subject did not end there; he also treated the subject in his academic courses.[19] English mathematical literature of the late nineteenth century was also rich in contributions on the subject. We can cite, in addition to the already mentioned Thompson and Tait, the names of George Darwin and James Jeans, and finally Horace Lamb, who devoted to the subject the last chapter of his famous treatise on hydrodynamics.[20]

During the whole of the nineteenth century, the gravity measurements, i.e., the measurements of the acceleration of gravity at ground level in various places on the terrestrial surface, as we have recalled in the previous section, were executed with the use of the pendulum. This method was preferred to the other, as theoretical as possible, of using the equation of the free fall of a body and making observations of its motion, since it was difficult to make accurate observations of the motion of bodies whose velocity has an increase of about 980 cm per second. In contrast, in the twentieth century, the improvement of technology will make the latter method preferable, but we shall not go into this. Instead let us see how the pendulum was used.

As we know, if we refer to the mathematical pendulum (a point mass m suspended on a weightless wire of length l), in the case of small oscillations the period is given by $T = 2\pi\sqrt{\frac{l}{g}}$, where g is the acceleration of gravity. Therefore if one wants to find g, l being known, it is necessary to measure T. This can be done by measuring its inverse, i.e., by counting the number of oscillations per second. But, in order for a measure of this sort to give trustworthy results, it is necessary that the actual pendulum used in practice resemble as far as possible the mathematical pendulum. The necessary elements for being able to reproduce a mathematical pendulum are the centre of mass and of the centre of oscillation. This was overcome

---

[17]H. Poincaré: "Sur l'équilibre d'une masse fluide animée d'un mouvement de Rotation". *Comptes rendus hebdomadaires de l'Académie des sciences de Paris* 100 (1885): 346-348; quote on p. 348. With the same title, he published another communication in *Acta Math.* VII, 259 (1885).

[18]F. Tisserand: *Traité de Méchanique Céleste*, 4 vols. (Paris, Gauthier-Villars, 1889–96), vol. II, p. 108.

[19]See *Figures d'équilibre d'une masse fluide: Leçons professée à la Sorbonne en 1900* par H. Poincaré – redigée par L. Dreyfus (Paris: Gauthier-Villars, 1902).

[20]*Hydrodynamics by Sir Horace Lamb* (1st. ed. 1879, 6th ed. 1932) (Cambridge University Press; rpt. Dover, 1997).

by Henry Kater (1777–1835), a British physicist and army captain, who in 1817 invented the reversible pendulum. The reversible pendulum, by having two centres of rotation (by which after a corresponding adjustment it is possible to achieve the same oscillation time), made the determination of the centre of mass and of the centre of oscillation unnecessary.

As said above, the pendulum method is no longer applied but it governed the measurements of gravity during the whole of the nineteenth century.

The measurement of g is fundamental for defining, at any given place, the surface of the geoid. The geoid's surface coincides with mean sea level over the ocean and continues in continental areas as an imaginary sea-level defined by spirit level. Thus it is perpendicular, in any point, to the physical plumb line and approximates locally, the shape of a regular oblate spheroid.

Mathematically speaking, the geoid is an equipotential surface, i.e., over its entire extent the potential function is constant. The potential of which we are speaking is the potential of the terrestrial gravitational field (it has been proven that the terrestrial gravitational field is a conservative field so it admits a potential) whose lines of force are by definition perpendicular to the equipotential surface and then tangent at the points of the geoid's surface to the local vertical (the plumb line).

If we assume as a reference system a Cartesian system having its origin in the centre of mass of the Earth and its z-axis along the axis of terrestrial rotation, we can in this reference system give the expression of the force acting on a mass point at the surface of the geoid and arrive successively to the equation of the geoid.

We shall make the following simplifying assumptions:

- we consider the Earth as a rigid body;
- we assume that the motion of rotation of the Earth around its axis takes place with constant angular velocity (ω);
- we assume that the motion of revolution of the Earth around the Sun is locally similar to a translation (uniform rectilinear motion);
- we neglect any other motion.

On these assumptions, the force acting on a point of unit mass, and then the total gravitational acceleration, will be composed of two terms: the acceleration due to the Newtonian attraction of the whole terrestrial mass and that due to the rotation of the Earth assumed to take place with angular velocity ω. Therefore we shall have the vector representing this total acceleration given by

$$\overline{g} \equiv \begin{cases} g_x = X + \omega^2 x \\ g_y = Y + \omega^2 Y \\ g_z = Z \end{cases}$$

In a conservative field, the force is the gradient of the potential and so is in this case.

The three components X, Y, Z, due to Newtonian attraction, will be obtained as the components of the gradient of the Newtonian potential

$$U = \int_E \frac{fdm}{r} = \int_E f \frac{\rho dv}{r},$$

where dm indicates the mass element and ρ is the density of the Earth, r the distance from the centre and the integral is extended to the whole volume of the Earth.

As the components due to rotation, they are obtained from the potential of the centrifugal force

$$U_1 = \frac{\omega^2}{2}(x^2 + y^2).$$

Therefore the equation of the equipotential surface (and of the geoid) will be

$$W = U + U_1 = \int_E f \frac{\rho dv}{r} + \frac{\omega^2}{2}(x^2 + y^2) = C,$$

where W is the total potential and C is a constant which will depend on the conventions with which the mean sea level will have been characterized.

Since the density $\rho = \rho(x, y, z)$ is not a known function (because we do not know the distribution of the density inside the Earth), the equation of the geoid cannot be expressed in closed form. On that account recourse has been made to a convergent series expansion of the potential, starting from the spherical harmonic expansion of the solution of the Laplace equation $\Delta U = 0$.[21] The geoid's surface is used as a reference surface for measuring heights and depths on the Earth.[22]

It seems to us that, from the general point of view, i.e., that of the principles which have inspired the measurements on the Earth, there are no substantial novelties, save in methodologies suggested by the technological developments in geodesic surveying up to the end of the first half of the twentieth century, that is, until the entrance of the artificial satellites. Until that time, man had measured the Earth by walking on it; afterwards, he started to look at the Earth from outside and from on high: a new horizon was wide open.

As is well known, the era (if we can call it that) of artificial satellites began in 1957 with the launch of Sputnik by the Soviet Union. Already in the years immediately following various missions on the part of the two leading powers (USA and USSR) led to the accurate determination of the shape of the geoid.

---

[21]We cannot develop this subject here, and refer the interested reader to the textbooks of mathematical physics. Well suited in this case is also the classical textbook by W. Torge: *Geodesy*, 3rd rev. ed. (Berlin & New York: 2001).

[22]For this for this particular argument we again refer to W. Torge: *Geodesy*, op. cit.

Since then artificial satellites have been used for the scientific research, communications, military purposes and more, and, in the case of communications, we can say that they have come to be part of our everyday lives.

The vastness of the applications, both for the scientific research and for the other purposes, is such that usually one classifies the techniques used by instrument platforms in terms of a “spatial” vocabulary: Earth-to-space methods; space-to-Earth methods; space-to-space methods.

Our “journey”, as the reader will have understood, substantially came to an end before the advent of artificial satellites. And therefore we close our account here.[23]

Suggested Readings

Alexander Ross Clarke: *Geodesy*. Oxford: Clarendon Press, 1880. This is the book of the British geodesist (1828–1914), quoted in §6.1, which was the first major survey of the subject for many years.

Kaula, W. M. (1996). *Theory of the satellite geodesy: Applications of satellites to geodesy*. Blaisdell Publishing Company. Dover reprint, 2000.

Sansò, F., Sideris M. G. (Eds.) (2013). *Geoid determination: Theory and methods*. Berlin: Springer. For consultation by a reader with a mathematical background.

Smith J. R. (1997). *Introduction to geodesy: The history and concepts of modern geodesy*. Hoboken: Wiley. A comprehensive and accessible book.

[23]On satellite geodesy, see G. Seeber: *Satellite Geodesy* (Berlin & New York: Walter de Gruyter, 2003).

Epilogue. Impressions of a Journey

If our "journey" through about twenty-five centuries has told us anything, it is that an idea (in our case the idea that the Earth has a spherical shape) requires suitable conditions and due time in order to establish itself and become a shared inheritance. Some might think that this is obvious, and it certainly is so, but there are cases which are baffling.

The case which we find most surprising is that of the Roman world. In the time of the Republic and of the Empire—we can say about seven centuries—the idea of the Earth's sphericity apparently never became popular. Evidence of this is provided by the fact that even among the intellectuals, as we have seen, the opinion was controversial, and then it was impossible to have a common opinion at the popular level. At a certain point, the entrance of the Christian Church and the consequent interpretation of the Bible was a further obstacle to the diffusion of the idea.

More evidence can also be found in the fact that the Romans, the greatest road constructors of Antiquity, were never enticed by the idea of "measuring the world". The oecumene known at that time was in large part under Roman domination and certainly the distances between the most important districts had been measured; this was sufficient for the public administration. One has the impression that the Roman people moved around without wondering if the Earth was flat or not.

However, Vitruvius, writing at the time of Augustus, in his book *De Architectura*, in the chapter where he deals with various properties of waters, notes that:

> Perhaps some reader of the works of Archimedes will say that there can be no true levelling by means of water, because he holds that water has not a level surface, but is of a spherical form, having its centre at the centre of the earth.[1]

[1]Vitruvius, *De Architectura* VIII, 3–5 (Eng. trans. M. H. Morris).

D. Boccaletti, *The Shape and Size of the Earth*,
https://doi.org/10.1007/978-3-319-90593-8

With regard to the longest leg of our journey, which is, according to our metaphor, the one travelled slowest, we wish to add a couple of considerations to what has already been said at the end of Chap. 3. While in the Arab countries, starting from the eighth century, the world of learning was flourishing, in Europe learning limped along inside the monasteries starting from what survived of Latin erudition. What we wish to emphasize is that at that time (at least from the cultural point of view) it was as if there were two parallel, non-communicating worlds. This situation lasted until when, in Muslim-dominated Spain, there was the encounter of Arabs with Western intellectuals. This fact may surprise someone who, like ourselves, lives in the world of instant communications.

Another consideration concerns the survival of Pythagorean philosophy in the West. As we know, the transmission into the West of Pythagorean philosophy occurred, in the Roman world in the time of emperors, by means of the biographies of Pythagoras written by Diogenes Laertius (180–240 AD), Porphyrius (233–305 AD) and Iamblicus (245–325 AD). But one notes the absence of Pythagorean philosophy *per se* in the Middle Ages, where the influence of Neoplatonic philosophy (also tied to the Pythagorean tradition) seems to refer strictly to some of Plato's works. There was a revival of Pythagorean philosophy at the beginning of the Renaissance, but this revival only appertained to what we might call the "magical" form of Pythagorean philosophy, tied to number theory, and not to the cosmological model.[2] Only Copernicus, in the dedicatory letter to Pope Paul III that prefaces his *De Revolutionibus*, wanting to refer himself to the ancient philosophers, quotes "Philolaus the Pythagorean",[3] even if not Pythagoras himself.

Reconsidering the Middle Ages, one may be astonished that it took such a long time for the idea of the sphericity of the Earth to become accepted at the level of the intellectual debate. However, one must consider that the thinkers who dealt with that problem, at least in a large part of this period, belonged to religious orders and so also dealt with theology (for instance, the problematic insertion of the term *filioque* in the Nicene Creed); the nature of the Earth only came in because in Genesis it appears in the narration of the creation. Nevertheless, ideas which presaged a scientific conception were also incubating in that period (see Roger Bacon).

At the beginning of the Modern Era, between the fifteenth and the sixteenth centuries, the experiences of the navigators swept away the bookish culture of the preceding centuries and finally drove the interested scholars "to get their hands dirty", projecting and executing measurements on the ground: modern science took its first steps.

The first of the new scholars, as we have seen, was Jean Fernel, who shifted from theoretical speculations to effective measurements by combining the indispensable astronomical observations with the calculation of the path on the ground inventing

---

[2]An appealing monograph on this subject is Paolo Casini: *L'antica Sapienza italica. Cronistoria di un mito* (Bologna, Il Mulino, 1998).

[3]See: "Dedication of the Revolutions of the Heavenly Bodies by Nicholas Copernicus (1543) to Pope Paul III", in *The Harvard Classics: Prefaces and Prologues to Famous Books* (New York: P. F. Collier and Son, 1938), pp. 52–57; reference to Philolaus on p. 55.

an ad hoc experiment. In this way Fernel founded modern geodesy. We note that in the same year (1528) in which he published his *Cosmotheoria*, Fernel also published a short work, *De Proportionibus*,[4] reprising a subject held dear by the philosophers of Merton College (Oxford) in the fourteenth century. After his exploits as a surveyor, Fernel devoted himself to medicine and obtained results so important that he is considered the father of the modern physiology.

If Fernel was the first, in the Modern Era, to measure the arc of meridian and consequently the circumference of the Earth, about ninety years later Willebrord Snell would be the first to measure the arc by means of a new technique that will also be used in the future: triangulation. But we have already spoken of this. We would instead like to emphasize that Snell was the first to point out the problem of the different units of measurement in the different countries and thus of the difficulty of comparing results of measurements among researchers using different units. This is the first time that the problem is put forward and so extensively (Snell dedicates five chapters of his book to it). He anticipates, at least in his sensitivity towards the problem, the great mathematicians of the Académie des Sciences who will "create" the metre and the decimal metric system.

As to these last, what excites our astonishment and admiration is the dedication with which the greatest French mathematicians and physicists applied themselves to the execution, both practical and theoretical, of the measurement of the "Méridienne de Paris". As is natural, there was the presence of the national pride, even in turbulent times with the Revolution under way, and the rivalry between the Académie and the English Royal Society, but the effort they lavished on the endeavour was extraordinary. Even D'Alembert, "mathematician, man of letters",[5] the prophet of rational mechanics, was involved and supported Maupertuis' expedition, besides contributing to the mathematical theory of the figure of the Earth.

In the nineteenth century, the activism of the French scientists "infected" their European colleagues, and also a great (nay, the greatest of the nineteenth century) mathematician: Carl Friedrich Gauss. Gauss, in 1823, accomplished the measurement of the Hannover arc. For him this was also the opportunity for coupling the study of the figure of the Earth with his researches on differential geometry which resulted in his famous work.[6] We are pleased to note that, just as a novel has been written about the measurements of the French mathematicians, a recent novel has been dedicated to the work of Gauss as well.[7] In our opinion, the measurement effected by Gauss, although commissioned by the Kingdom of Hannover, was the

---

[4]J. Fernel, *Joannis Fernelii Ambianatis de proportionibus Libri duo* Paris: Ex aedibus Simonis Colinaei, 1528.

[5]So he was described by Louis de Broglie in a commemoration, in 1952, of the bicentennial of the *Encyclopédie*; see Louis de Broglie, "Un Mathematicien, Homme de Lettres", *Revue d'histoire des sciences* Année 1951 4(3–4): 204–212.

[6]C. F. Gauss: *Disquisitiones generales circa superficies curvas* (Göttingen: Typis Dieterichianis, 1828).

[7]Daniel Kehlmann: *Measuring the World* (Vintage Books, 2007) (First German edition, 2005).

last one that we can insert into the tradition inaugurated by Eratosthenes of the scientists (mathematicians and astronomers) fascinated by the idea of "measuring the world".

As we have seen, in the 1880s a new discipline was born: geodesy. From then on triangulations were no longer an activity of scientists who sought to state definitively which shape the terrestrial surface might have, but an activity practised by a particular category of scientists, who move within a well-defined field that typifies their research. The advent of artificial satellites has completely replaced earlier observational techniques and a new powerful technology has been invented.

Author Index

D. Boccaletti, *The Shape and Size of the Earth*,
https://doi.org/10.1007/978-3-319-90593-8

The manufacturer's authorised representative in the EU is Springer Nature Customer Service Centre GmbH, Europaplatz 3, 69115 Heidelberg, Germany. If you have any concerns regarding our products, please contact ProductSafety@springernature.com

Printed and bound by CPI Group (UK) Ltd Croydon, CR0 4YY
30/09/2026
02225655-0002